walkermaths 1.3

Statistical and Mathematical Interpretation

Charlotte Walker and Victoria Walker

Walker Maths: 1.3 Statistical and Mathematical Interpretation
1st Edition
Charlotte Walker
Victoria Walker

Cover designer: Cheryl Smith, Macarn Design
Text designer: Cheryl Smith, Macarn Design
Production Controller: Siew Han Ong

Any URLs contained in this publication were checked for currency during the production process. Note, however, that the publisher cannot vouch for the ongoing currency of URLs.

Acknowledgements
The authors and publisher wish to thank the following people and organisations for permission to use the following resources in this workbook.
StatsNZ for data on pages 19 (Q11), 27 (Q3), 28, 32 (top and Q1), 33 (Q2 and 3), 47, 48, 62 (Q7), 70, 71, 86, and 99. Ministry of Health for data on pages 26 (Q3), 27 (Q4), and 61 (Q5). Figure NZ for data on pages 29 (Q6), 34 (Q5), 58 (Q5), 61 (Q4), 62 (Q6), 66, 67, 68, 72, 74, 75, 76 (Q1), 77, 78, 79, 80, 83, 85, 87, 88, 89 (Q3), and 90. Phillipa Agnew for the Oamaru penguin data on pages 39 and 84. Ministry of Social Development for data on pages 45, 46 and 91. Companion Animals in New Zealand for data on pages 49 and 58 (Q4). New Zealand Police for data on pages 34, 50, 73 and 92. ACC for data on pages 60 (Q2), 94, and 137. Ministry for the Environment for data on page 69. Consumer for data on pages 76 (Q2) and 141 (Resource B). Environment Science & Research for data on page 82. Reserve Bank of New Zealand for data on page 93. Electoral Commission for data on page 100 (Q8). The Ministry of Foreign Affairs and Trade for data on page 101 (Q10). Office of the Prime Minister's Chief Science for data on page 141 (Resource A). Rabobank and KiwiHarvest for data on pages 141 (Resource C), 143 (Resources D, E and F), and 145 (Resource G).
All other images are courtesy of Shutterstock and iStock.

We also wish to acknowledge the trusted kindred spirits throughout the country for your help in preparing this title.

For product information and technology assistance,
in Australia call **1300 790 853**;
in New Zealand call **0800 449 725**

For permission to use material from this text or product, please email
aust.permissions@cengage.com

National Library of New Zealand Cataloguing-in-Publication Data
A catalogue record for this book is available from the National Library of New Zealand.

978 0 17 047755 0

Cengage Learning Australia
Level 5, 80 Dorcas Street
Southbank VIC 3006 Australia

Printed in China by 1010 Printing International Limited.
2 3 4 5 6 7 28 27 26 25 24

 This icon appears throughout the workbook. It indicates that there is a worksheet available which has been collated from the original Level 1 WalkerMaths series. The worksheets can be accessed via the Teacher Resource for this workbook (available for purchase from nz.sales@cengage.com). These worksheets are an additional resource which can be used to support your students throughout the teaching of this standard.

CONTENTS

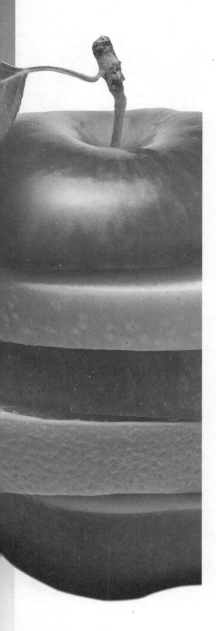

   ISBN: 9780170477550

Statistical investigations

Census and sample

- The first decision that has to be made in any investigation is whether to take a census or a sample.

A census compared with a sample

Census
- You collect data from **every member of the population**.
- You get very accurate information.
- However, it is often impossible and usually very expensive to do.
- It is best to do a census if the answer to the question **really matters**.
- In New Zealand, a census of the population takes place every five years.

Sample
- You collect data from **just some of the population**.
- You don't get such accurate information.
- However, it is much easier and cheaper to take a sample.
- It is important that the sample is selected **fairly**.

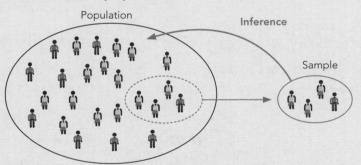

The results from the sample provide an **estimate** or **approximation** about the population. This is known as an **inference**.

Which would be more appropriate for getting information for the following situations, and why?

	Question	Census or sample?	Why?
1	What do New Zealanders think should be the bird of the year?		
2	Do New Zealanders want a new flag?		
3	What is the favourite crisp flavour of New Zealanders?		

Sampling and bias

- It's important to consider whether a sample is likely to reflect the characteristics of the population.

Selecting a sample which reflects the population
- Bias occurs when some members of the population are more likely than others to be selected for the sample, so the sample does not truly represent the population.
- There are sampling methods that make it more likely that the sample will be representative of the population.

Sample size
- Each sample needs to be large enough to ensure that it reflects the characteristics of the population.
- The larger the sample size, the more precise the population estimates are likely to be.
- The proportion of any subgroup in the sample should roughly reflect its proportion in the population.
- When comparing two samples, their sizes **do not** have to be equal.

Answer the following questions.

1 a The numbers of black dogs and white dogs in this population are roughly equal. Count the numbers of black dogs and white dogs in each sample.

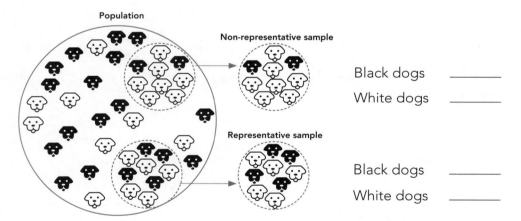

Black dogs   _____

White dogs   _____

Black dogs   _____

White dogs   _____

  b The first sample is non-representative because the numbers in the sample are/are not proportional to those in the population.

2 Discuss with your classmates whether the sample is likely to be representative or non-representative in these situations.

  a Population: All the apples on a tree.
Sampling method: Shaking the tree and collecting those that fall off.

  b Population: All the students in a school.
Sampling method: Questioning the first 100 students through the gate in the morning.

  c Population: The New Zealand population over 15 years old.
Sampling method: Ringing people on their cellphones between 8 am and 6 pm.

  d Population: Kauri trees in a forest.
Sampling method: Sampling all the trees within 10 m of a track.

Sampling methods

1 Random sampling methods

These are random methods which, if executed well, are likely to produce a representative sample.

Simple random	Randomly select participants from the population.	
Stratified	Divide the population into **strata** (groups) and take a **simple random sample from each stratum** (group).	
Cluster	Randomly select groups from within the population (clusters), then analyse/survey every member of the selected clusters. Each cluster must be representative of the population.	
Systematic	Choose a **random starting point** and then sample members of the population at a **regular interval**.	

2 Biased sampling methods

These methods are likely to produce a sample that is NOT representative of the population.

Volunteer or self-selection	People who volunteer often have different characteristics to non-volunteers. For example, to volunteer to donate blood, people have to be 16 years old, be in good health, etc.	
Convenience	Participants who are close at hand are selected.	

Identify the type of sampling in each of these situations. In some of these, it may be appropriate to select more than one type. Justify your choice. State whether the sample is likely to be representative or non-representative.

1   Population: Shoppers in a mall.
     Sample: People at the mall supermarket were asked to fill in a questionnaire.

    a   ☐ Simple random   ☐ Cluster   ☐ Self-selected
    ·   ☐ Stratified   ☐ Systematic   ☐ Convenience

    Justification: _____

    _____

    b   This is likely to produce a representative/non-representative sample.

    Reason: _____

    _____

2   Population: All the employees at a company.
     Sample: All the employees were each given an ice-cream stick and asked to write their name on it. These sticks were put together in a container, shaken up, and then five sticks were chosen.

    a   ☐ Simple random   ☐ Cluster   ☐ Self-selected
        ☐ Stratified   ☐ Systematic   ☐ Convenience

    Justification: _____

    _____

    b   This is likely to produce a representative/non-representative sample.

    Reason: _____

    _____

3   Population: Ākonga (students) in a PE class in which 18 played volleyball and 9 played football.
Sample: Of the volleyball players, six were randomly selected. Of the football players, three were selected.

a   ☐ Simple random   ☐ Cluster   ☐ Self-selected
        ☐ Stratified   ☐ Systematic   ☐ Convenience

Justification: _____

_____

b   This is likely to produce a representative/non-representative sample.

Reason: _____

_____

4   Population: All students in a school.
Sample: The two vertical (years 9–13) form classes nearest to the administration block.

a   ☐ Simple random   ☐ Cluster   ☐ Self-selected
        ☐ Stratified   ☐ Systematic   ☐ Convenience

Justification: _____

_____

b   This is likely to produce a representative/non-representative sample.

Reason: _____

_____

5   Population: All the students in the school.
Sample: Starting from the eighth student, every tenth person on the school roll.

a   ☐ Simple random   ☐ Cluster   ☐ Self-selected
        ☐ Stratified   ☐ Systematic   ☐ Convenience

Justification: _____

_____

b   This is likely to produce a representative/non-representative sample.

Reason: _____

_____

6   Population: All the students in the school.
Sample: All those who want stay behind after assembly to answer a questionnaire.

a   ☐ Simple random   ☐ Cluster   ☐ Self-selected
        ☐ Stratified   ☐ Systematic   ☐ Convenience

Justification: _____

_____

b   This is likely to produce a representative/non-representative sample.

Reason: _____

_____

Collection of data

- When information is provided in an article, report, infographic or other display, it is important to evaluate the data collection process (methodology), as this affects the accuracy, validity and reliability of the results.

When collecting data, measurements should be:

1 **Accurate:** the data values are close to the true value.
If a dog is being weighed, the scales should calibrated (checked that they are not weighing consistently too high or low), and the dog should be sitting still.
2 **Valid:** the data values actually represent what they claim to.
When the mass of a dog is required, make sure the dog is dry and its collar is removed while it is being weighed.
3 **Reliable:** measurements are collected consistently, so different investigators get similar results.
4 **Collected ethically.**

Example: Measurements of people's hand spans were being collected. Explain how the reliability of the data could be affected by the use of different methods or instructions.

1 People must be told to spread their fingers as wide as possible.
2 The measurement needs to be made when the entire hand is spread and pressed down onto a flat surface, not measured in the air.

If the same instructions are followed by everyone, the data collection method will be consistent, making the results more reliable.

Answer the following questions regarding the measurement of people's heights.

1 Explain how removing shoes and flattening hair when measuring a person's height increases the validity of the data recorded.

_____

_____

2 Explain how using the same scale and ensuring the height is read at right angles to the scale increase the accuracy of the data recorded.

_____

_____

_____

3 Explain how having a set of clear instructions for those measuring people's heights would increase the reliability of the data.

_____

_____

Data collection processes

There are three types of data collection processes.

1 Surveys and **polls**
- A **questionnaire** is the set of questions used in a poll or survey.
- A **survey** is the process of collecting data using a **questionnaire**.
- A **poll** is when people are asked **one question** about their opinion.
- If the data collection process was **well-designed** and there is **enough evidence** in the data, a sample-to-population **inference** can be made.

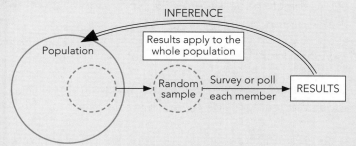

2 Observational studies where people or objects are **observed** without altering or controlling their behaviour in any way.
- Data is collected using one of the **sampling methods**.
- If the data collection process was **well designed** and there is **enough evidence** in the data, a sample-to-population **inference** can be made.

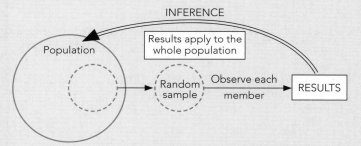

3 Experimental studies are when the investigator applies a treatment to one half of the participants and not to the other half (the control group).
- It is essential that the participants are **randomly allocated** (e.g. by tossing a coin) to the groups.
- Participants do not know which group they have been allocated to.
- One group has no treatment (the control) and the treatment is applied to the other group (or groups).
- If the experiment was well-designed and there is **enough evidence** for a difference between the groups, then a claim can be made that **the difference between the groups was caused by the treatment**.

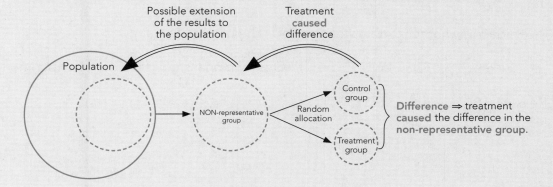

Examples: For each of the questions below, identify and justify what type of data collection process was used.

1 Ākonga were given a form to fill in, which included a series of questions about the hobbies and activities that they enjoy.

   This is a survey, because ākonga were answering several questions on a questionnaire.

2 Teenagers were asked what they think is the best pizza topping.

   This is a poll, because the teenagers were asked one single question which asked for their opinion.

3 Data was collected on which drinks ākonga choose at the formal.

   This is an observational study, as the investigator has no control over what drinks ākonga choose and can only observe what each ākonga chooses.

4 An investigation was carried out to see if putting fertiliser on plants affects the growth of plants. One group had fertiliser and one group did not.

   This is an experimental study, as the investigator controls whether fertiliser is put on each plant.

For each of the following situations, identify which data collection process is described and justify your choice.

1 Kaiako (teachers) were asked a series of questions about ākonga in their classes.

   ☐ Survey   ☐ Poll   ☐ Observational study   ☐ Experimental study

   Reason: _____

   _____

2 Students were asked who their favourite All Black was.

   ☐ Survey   ☐ Poll   ☐ Observational study   ☐ Experimental study

   Reason: _____

   _____

3 Investigating whether extra watering affects the height of sunflowers.

   ☐ Survey   ☐ Poll   ☐ Observational study   ☐ Experimental study

   Reason: _____

   _____

4 Questions were given to employees at a company, asking about their experience and skills.

☐ Survey   ☐ Poll   ☐ Observational study   ☐ Experimental study

Reason: _____

_____

5 Investigating the results of sports games over the season.

☐ Survey   ☐ Poll   ☐ Observational study   ☐ Experimental study

Reason: _____

_____

6 Collecting data on the prices of fruit over the last 10 years.

☐ Survey   ☐ Poll   ☐ Observational study   ☐ Experimental study

Reason: _____

_____

7 Investigating who can run faster — ākonga wearing running shoes or those with bare feet.

☐ Survey   ☐ Poll   ☐ Observational study   ☐ Experimental study

Reason: _____

_____

8 Weighing students' backpacks and analysing the data.

☐ Survey   ☐ Poll   ☐ Observational study   ☐ Experimental study

Reason: _____

_____

9 An experiment is being performed on the class, so they need to be randomly allocated to two roughly equal groups. Describe two ways (other than tossing a coin) that they could be randomly allocated to each group.

Method 1: _____

_____

Method 2: _____

_____

Survey methods

A survey **method** is the process used in order to collect data obtained by questioning a sample of people.

Common **survey methods** for contacting subjects include:
- Distributing a pen-and-paper questionnaire
- Telephoning to get oral responses
- Arranging to meet subjects for a face-to-face interview
- Internet contact via a website
- Email contact with an online questionnaire.

When deciding which survey method(s) to use, consider some of these factors:
- the **time** it takes to prepare, collect and process the data ready for analysis
- the **cost** of collecting data
- the **sampling frame** (a list of people in the population from which a sample is selected)
- whether to collect data from a wide **geographical area** or a small local area
- what the **response rate** is likely to be
- possible **biases** present in the choice of method.

Example: The school council needs to survey opinions of senior students regarding the location, music, food, etc. for the school formal. They plan to put a pile of questionnaires in the school foyer, with a box into which the completed questionnaires can be placed. List some advantages and disadvantages of this plan.

Advantages:
- Cheap to do.
- Easy to do.

Disadvantages:
- Self-selected response means potential bias.
- Requiring responses to be named might restrict who responds and the responses obtained.
- Because the responses will be handwritten, collation is likely to take a lot of time.

Considerations when writing the questionnaire:
- Anonymous responses mean that anybody (e.g. junior students) could respond, and people could put in several responses, which would bias the results.
- The box for responses might not be secure, so responses could be removed.

A survey with 10 questions needs to be given to a sample of 30 people. For each situation below, discuss the advantages and disadvantages of each. **Hint:** Think about the cost of creating the surveys, and the time spent collecting and processing the data.

1   Delivering a survey to letterboxes of people living near a school, where people fill in the form using a pen/pencil and post the surveys using an included pre-paid envelope.

Advantages: _____

_____

_____

Disadvantages: _____

_____

_____

     ISBN: 9780170477550

2   A call centre dials cellphone numbers and questions the person who answers the call. The questions are listed on the screen for the interviewer, who then enters the answers using a drop-down menu.

Advantages: _____

_____

_____

Disadvantages: _____

_____

_____

3   The dean does face-to-face interviews of ākonga and records each interview. The conversation is then transcribed into a computer.

_____

_____

_____

_____

_____

_____

4   Parents go to a school's website to fill in an online survey, where they select their answers from a list or drop-down menu, or type in short-sentence answers.

_____

_____

_____

_____

_____

_____

5   A surveyor stands in a mall with a clipboard and asks passersby a series of questions.

_____

_____

_____

_____

_____

_____

Types of survey questions

- Statistics can be gathered by observation or by questioning.
- There are a number of types of survey questions that can be asked.
- It's important to choose the appropriate type for the data that you want to collect.

Type		Description	Example
Closed	Binary	Has only two answers.	Did you walk to school today? Yes ☐ No ☐
	Multi-choice	More than two possible answers but defined.	What colour eyes do you have? Brown ☐ Blue ☐ Green ☐ Other ☐
Open	Short	Requires a one- or two-word response.	What is your favourite ice-cream flavour?
	Long	Allows a sentence or paragraph answer.	What are your thoughts on compulsory participation sport in schools?

Closed questions are easier to analyse, but it's important to make sure they will give you the information you need to answer your investigative question.

Open questions can lead to any sort of response, so there needs to be a good reason for using them.

What types of question are the following? Select your answers from these: closed binary, closed multi-choice, open short and open long.

	Question	Type of survey question
1	What vegetable do you like best?	
2	Do you currently have any books that you have borrowed from the school library? ☐ Yes ☐ No	
3	How do you feel about the school uniform?	
4	What would you prefer for breakfast? Cereal ☐ Toast ☐ Eggs ☐	
5	What was the best thing you did during the last holidays?	

Ethics

- Data is a **taonga** (treasure), and sensitivity and respect is needed regarding data that is collected, recorded, stored, presented and published.
- Always keep in mind: '**Behind every data point is a person, a business, an iwi, or a community**' (Liz MacPherson in 2019 in her role of Government Statistician and Chief Executive of Statistics New Zealand).

Some ethical considerations are:

1 Manaakitanga/Reciprocity: showing respect, humility, kindness and honesty.
- Fully informed and voluntary consent has been given.
- Data about an individual person/company is collected directly from that individual.
- Risks including physical and emotional distress, embarrassment or harm are minimised.
- Social and cultural responsibility are considered throughout the process.

2 Whakapapa/Relationship: researchers establish a relationship with those communities from whom they collect the data, and details about the data such as:
- why the data was collected.
- who (or what) it was collected from
- descriptions of the variables, etc.
- how it was collected.

3 Kaitiakitanga/Guardianship: those who have the data are caretakers; they do not own it.
- Data is stored safely and securely (e.g. where is it stored, who has access, etc.).
- Personal information is kept private and confidential (e.g. an individual cannot be identified from the data set).
- Individual people/companies can access their information, update it, and withdraw consent at any time.
- People/companies know when and where their information is being used.
- There are appropriate safeguards against inappropriate sharing of the data.

For each of the situations below, select and justify whether it is ethically appropriate or not.

1   Taking information from a private message thread on Snapchat.

         ☐ Ethically appropriate   ☐ Not ethically appropriate

Reason: _____

_____

2   Publishing a graph of class results of an internal assessment for the whole class to see.

         ☐ Ethically appropriate   ☐ Not ethically appropriate

Reason: _____

_____

3 Asking someone to sign a consent form after their data was collected.

☐ Ethically appropriate ☐ Not ethically appropriate

Reason: _____

_____

4 Storing assessments in a locked cupboard.

☐ Ethically appropriate ☐ Not ethically appropriate

Reason: _____

_____

5 Providing data on a student's attendance to a potential employer without asking permission from the student.

☐ Ethically appropriate ☐ Not ethically appropriate

Reason: _____

_____

6 A participant information sheet being provided in multiple languages.

☐ Ethically appropriate ☐ Not ethically appropriate

Reason: _____

_____

7 Deleting data from the data set of a participant because their spelling was bad.

☐ Ethically appropriate ☐ Not ethically appropriate

Reason: _____

_____

8 Data was collected on books in the school library for the purpose of deciding what genres of books to buy next year. The data set was then used to identify students who do not like reading.

☐ Ethically appropriate ☐ Not ethically appropriate

Reason: _____

_____

9 The following is a sample from a data set.

Business name	Business type	Number of employees	Expenditure
Paper and More	Stationery	25	$354,000
Bits and Bobs	Fabric	12	$142,000
Lawn Care	Hardware	54	$651,000
…	…	…	…

Explain why it is inappropriate to be able to identify individual businesses in this data set.

_____

_____

_____

10 Explain how making a survey anonymous would affect the responses when collecting data on vaping among school students.

_____

_____

_____

11 Stats NZ Tatauranga Aotearoa collects data approximately every five years from all New Zealanders in order to have data that represents the population and helps to provide insights to inform policies and service provisions. The question(s) on gender and sex have changed between the 2018 and 2023 census:

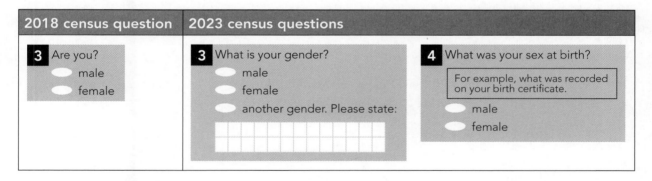

Explain why collecting data on both gender (a person's social and personal identity) and sex at birth is both more useful for data analysis and ethically appropriate.

_____

_____

_____

Putting it together

- Often we are not provided with the detailed methodology used during data collection.
- It is common practice to state the sample size, the source of the sample and possibly how the survey was conducted (e.g. online questionnaire).
- In addition, the population about which the inference is being made may not be clear.

Example: A new ice-cream flavour is tested in a local supermarket on a Monday between the hours of 10 am and 2 pm. Forty-two customers taste the ice cream and discuss their thoughts with the brand promoter.

Sample size adequate?	Sample representative of the population who might buy the ice cream?	Is the method of survey delivery acceptable?
☐ Yes ☒ No	☐ Yes ☒ No	☐ Yes ☒ No

Comments:
- The sample size of 42 is very small and unlikely to be sufficient to make a reliable inference about the new flavour, particularly as children are unlikely to be in the supermarket between 10 am and 2 pm.
- This sample is self-selected because customers have the choice to approach the brand promotor or not. Those who have time, like ice cream and like free samples are likely to participate.
- The survey is conducted at only one supermarket and at times when those who have normal working hours are unlikely to be sampled. It is also likely to exclude school-age students.
- When the customers discuss their thoughts about the new flavour, there is no guarantee that their opinions will be recorded accurately. It would be better to ask them to complete a form.

Answer the following questions.

1   A school with a role of 1500 wants to canvass opinions regarding a change of uniform. It sends a survey via its school email to 30 Year 13 students.

   a | Sample size adequate? | Sample representative of the population? | Method of survey delivery acceptable? |
   |---|---|---|
   | ☐ Yes
☐ No | ☐ Yes
☐ No | ☐ Yes
☐ No |

   b   Justify your selections.

   _____

   _____

   _____

   _____

2 A proposed change of road layout was posted on the social media account of the local council. Readers were invited to give their feedback about the changes in an online survey. The number who responded was 109.

a

Sample size adequate?	Sample representative of the population?	Method of survey delivery acceptable?
☐ Yes	☐ Yes	☐ Yes
☐ No	☐ No	☐ No
☐ Can't tell	☐ Unlikely	

b Justify your selections.

_____

_____

_____

_____

c Who should the target population include?

_____

_____

_____

d Suggest a better way of reaching the target population.

_____

_____

3 NZ On Air conducted a survey in April–May of 2023 and asked 1408 New Zealanders aged 15+ about the media they used 'yesterday'. Eight hundred and one interviews were completed by telephone using random-digit dialling, and 607 interviews were completed online in homes without a landline. Comment on any issues that might arise from this sample methodology and any assumptions you might make about it.

_____

_____

_____

_____

_____

_____

_____

Types of variables

There are three type of variables:

Categorical variables

These are descriptions or names.
Data recorded about each person/thing: **words**.
Typical question starts with 'What …'.
Examples: hair colour, brand of cellphone.

Discrete variables

These are **numbers** which are the result of **counting**.
Data recorded about each person/thing: **whole numbers**.
Typical question starts with 'How many …'.
Examples: number of siblings, number of shirts you own.

Continuous variables

These are **numbers** which are the result of **measuring**.
They can be **fractions** or **decimals**.
Typical question starts with 'How long/heavy …'.
Examples: height, weight, distance, time.

Write down which types of variables these are.

1 Favourite animal. _____

2 Number of pets. _____

3 Distance to school. _____

4 Eye colour. _____

5 Time it took you to travel home from school yesterday. _____

6 Number of people who live in your house. _____

Here is some data that has been collected:

	Question	Rangi	Tina	Koa
A	How tall are you?	166 cm	172 cm	145 cm
B	What is your favourite lolly?	Fruit bursts	Sour chews	Jelly beans
C	How long can you stand on one foot for?	34 s	1 min 48 s	7 s
D	How many brothers do you have?	2	0	4
E	What colour is your hair?	Black	Blonde	Brown

7 Which questions have answers that are categorical? _____

8 Which questions have answers that are discrete? _____

9 Which questions have answers that are continuous? _____

   ISBN: 9780170477550

Data display

- There are different ways of displaying data. Which is most appropriate depends on the **type** of variable.

	Numeric data	
Categorical	**Discrete data**	**Continuous data**
Tally chart Pictograph Dot plot Bar graph Strip graph Pie graph	Tally chart Pictograph Dot plot Bar graph Strip graph Line graph Scatter plot Box plot	Histogram Line graph/time series Scatter plot Box plot

Write the names of each graph, and the types of data for which each is appropriate.

1

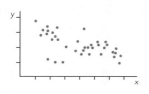

_____

_____

2

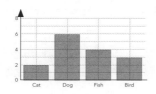

_____ Bar graph _____

_____ Categorical or discrete _____

3

_____

_____

4

_____

_____

5

_____

_____

6

_____

_____

7

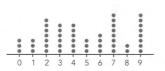

_____

_____

8

French	IIII
Te Reo	HHH IIII
German	II
Spanish	HHH

_____

_____

9

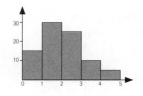

_____

_____

All sorts of other graphs are used, but remember, fancy doesn't necessarily mean better or easier to understand.

Graph summary

Pie graph

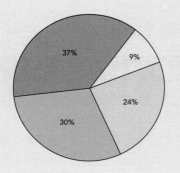

Data: Categorical or discrete numeric.

Advantages
Excellent for showing proportions of 100%.

Disadvantages
Not so useful when there are more than five divisions of the data.
Does not clearly show the order of numeric data.

Alternatives
Bar graph, dot plot, tally chart, pictograph.

Strip graph

Data: Categorical or discrete numeric.

Advantages
Shows order, so good for numeric data.
Excellent for showing proportions of 100%.
Excellent for comparing proportions in different groups.

Disadvantages
Not so useful when there are more than five divisions of the data.

Alternatives
Bar graph, dot plot, pie graph, pictograph, tally chart.

Bar graph

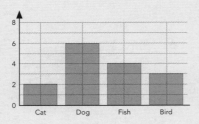

Data: Categorical or discrete numeric.

Advantages
Shows order, so good for numeric data.
Shows the shape of the distribution.
Can still work well where there are more than five divisions of the data.

Alternatives
Pie graph, dot plot, tally chart, pictograph.

Note: There are **gaps** between bars, and they are labelled **centrally** on the x-axis (or y-axis if the bars are horizontal).

Other name: Sometimes called a column graph.

Histogram

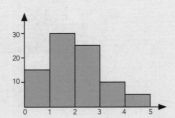

Data: Continuous.

Advantages
Shows order, so good for numeric data.
Shows the shape of the distribution.

Alternatives
Line graph.

Note: There are **no gaps** between bars, and labels are **between bars** on the x-axis.

 ISBN: 9780170477550

Line graph

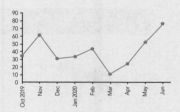

Data: Discrete or continuous numeric.

Advantages
Shows order of data.
Excellent for time series data.

Alternatives
Bar graph (discrete), histogram (continuous).

Dot plot

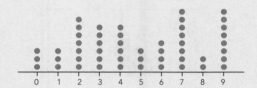

Data: Categorical or discrete numeric.

Advantages
Shows order, so good for numeric data.
Shows the shape of the distribution.
No loss of information: the original data can be reconstructed from a dot plot.
Shows unusual points or groupings.
Possible to calculate statistics (median, etc.) from a dot plot.

Alternatives
Categorical data: bar graph, tally chart, pictograph.
Discrete numeric: bar graph, box plot, tally chart, pictograph.

Box plot

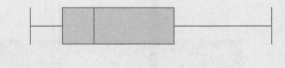

Data: Discrete or continuous numeric.

Advantages
Shows the symmetry of the distribution.
Shows values of median, quartiles and highest and lowest values.
Pairs of box plots are excellent for comparing distributions.

Alternatives
Discrete data: dot plot, bar graph, tally chart, pictograph.
Continuous data: line graph, histogram.

Other name: Sometimes called a box and whisker plot.

Scatter plot

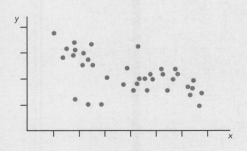

Data: Discrete or continuous numeric.

Advantages
Excellent for showing the relationship between two sets of data.

Alternatives
None.

Other name: Sometimes called a scatter graph.

Answer the following questions.

1   The members of a class were asked which pet they would most like to own. The results are shown in the table.

   a   What sort of data is this? _____

   b   Which of these graphs could you use to display this information?

   ☐   Bar graph

   ☐   Scatter plot

   ☐   Pie graph

   ☐   Line graph

   ☐   Histogram

   ☐   Dot plot

   ☐   Box plot

Pet	Frequency
Dog	11
Fish	4
Cat	8
Rabbit	6
Bird	1

2   The Ministry of Health did a survey in 2022 about the dietary habits of New Zealanders.

   a   What sort of data is this? _____

   b   How else could this data have been displayed?

   ☐   Bar graph(s)

   ☐   Scatter plot(s)

   ☐   Pie graph(s)

   c   How would you describe an infrequent red meat eater?

   Somebody who eats meat _____

   _____

Prevalence of eating red meat in people of 15 years or older	
Female	%
Never or infrequent	9.0
Once or twice/week	44.4
3+ times/week	46.6
Male	%
Never or infrequent	5.5
Once or twice/week	40.6
3+ times/week	53.9

ISBN: 9780170477550

3 This graph shows the ethnicities of New Zealanders from the 2018 census.

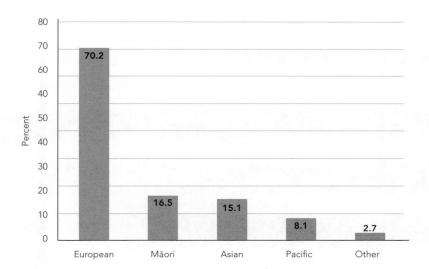

a Name another type of graph that would be appropriate for this data.

_____

b This data is not suitable data for a pie graph. Why?

_____

c What does this tell you about how the data was obtained?

_____

4 This graphic shows the number of female and male employees at one of the government ministries.

a What kind of graph is this?

☐   Line graph

☐   Bar graph

☐   Box plot

b What could have been done to improve this graph?

_____

_____

Gender: percentage

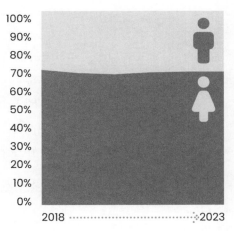

c Why might the content of this graph be considered inappropriate?

_____

5 Based on the 2018 census, Auckland Transport (AT) compiled this information on ways people travel to work or their education facility.

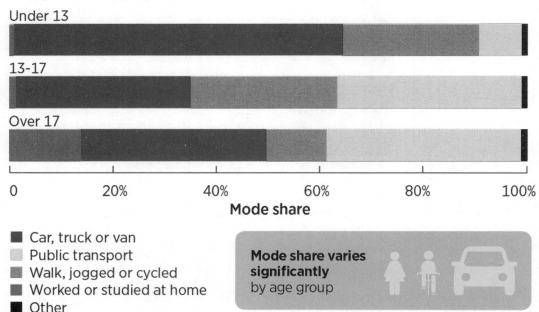

Mode share by Age group
Categories for age groups

Legend:
- Car, truck or van
- Public transport
- Walk, jogged or cycled
- Worked or studied at home
- Other

Mode share varies significantly by age group

a Which age group is most likely to use public transport? _____

b Which age group is least likely to use public transport? _____

c How else could this data have been displayed?

☐ Bar graph(s)

☐ Scatter plot(s)

☐ Pie graph(s)

d Why would AT choose to display the information in this way?

☐ To be able to read exact percentages.

☐ To give a visual representation of the different modes of transport for each age group.

☐ To suggest that people over 17 years old are more concerned about the environment than those under 13 years old.

6 These graphs both display the reasons for student suspensions during 2021.

Pie graph

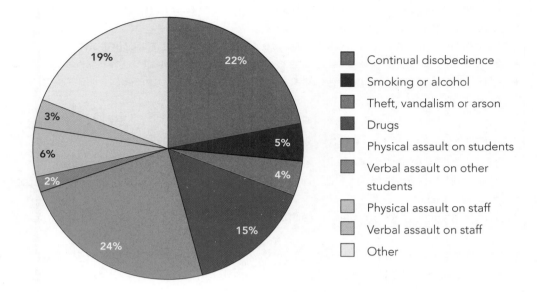

Bar graph
(with horizontal bars)

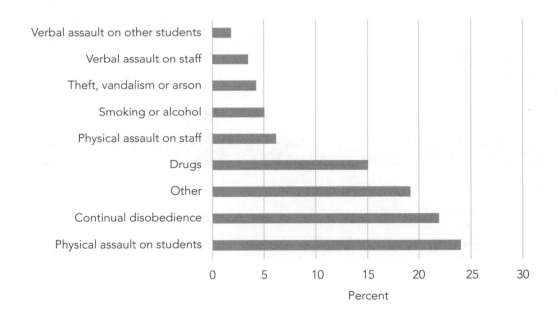

Which graph is the better display of this information? Explain your answer.

_____

_____

_____

_____

Line graphs

- Line graphs are often used to show changes in **discrete** or **continuous** numeric data.
- Straight lines connect the plotted points.
- When time is on the *x*-axis, these are called **time series** graphs.
- **Single** line graphs can be used to plot the data obtained from answers to a **summative** question.
- **Multiple** line graphs can be used to plot the data obtained from answers to a **comparative** question.

Time series

- Time series is a series of values collected at **regular** intervals of time.
- The intervals could be years, quarters, months, weeks, days, hours, minutes, seconds or even milliseconds.
- These are plotted on a line graph using a graphing program, with time on the *x*-axis.

Quarterly data: This means data points are given for every quarter throughout the year.
  Quarter 1 (Q1) is data from the start of January, February and to the end of March.
  Quarter 2 (Q2) is data from the start of April, May and to the end of June.
  Quarter 3 (Q3) is data from the start of July, August and to the end of September.
  Quarter 4 (Q4) is data from the start of October, November and to the end of December.

Monthly data: Unless you are told otherwise, M01 is January.

Daily data: Unless you are told otherwise, D01 is Monday.

Features of time series
Pattern
- There may or may not be a clear pattern.
- A repeating pattern usually occurs at regular intervals and has peaks and troughs.

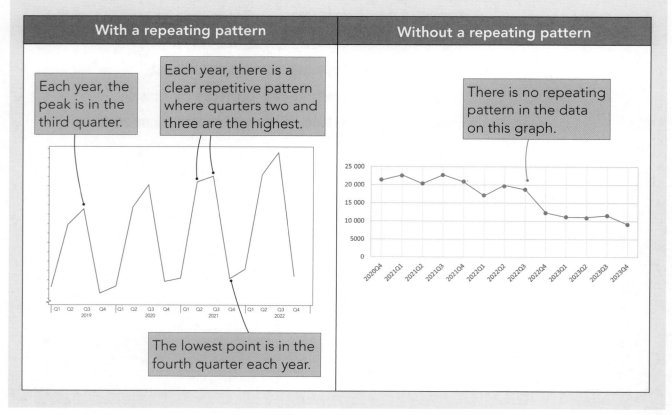

With a repeating pattern	Without a repeating pattern
Each year, the peak is in the third quarter.	There is no repeating pattern in the data on this graph.
Each year, there is a clear repetitive pattern where quarters two and three are the highest.	
The lowest point is in the fourth quarter each year.	

Trend

- A trend line can be added.
- This shows overall change.
- It could be increasing, decreasing, stable or inconsistent.

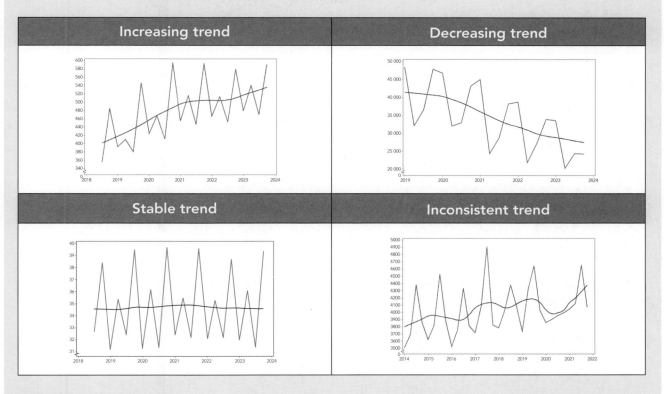

Variation

- This refers to the distance between the data and the trend line.
- This may increase, decrease, be consistent or be irregular.

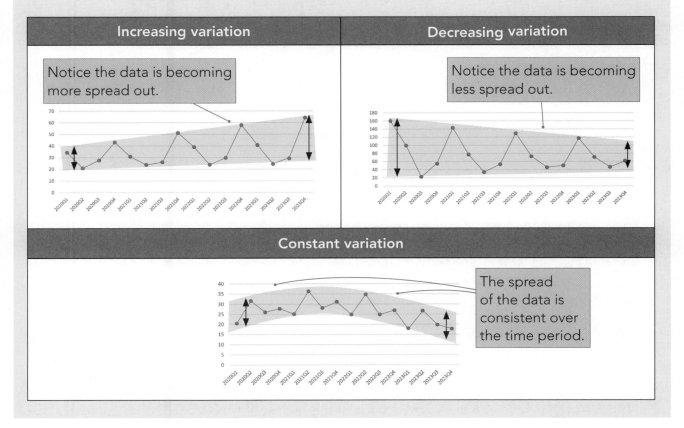

Example: The graph shows the monthly price per kg of apples in New Zealand over a period of three years. Describe the features of this graph.

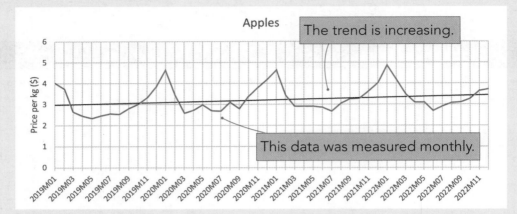

There is a clear repeating pattern in the price of apples per kg in New Zealand. The peak occurs in January and the lowest points occur throughout the winter months (M05–M07).

> Pattern + description

Over this period of time, there has been a slight increase in the price of apples from an average of approximately $3 in early 2019 to around $3.50 in late 2022.

> Trend + description

The variation in price is fairly consistent at around $2.50 between the peak and trough.

> Variation + description

1 This graph shows the cost of power for a New Zealand household.

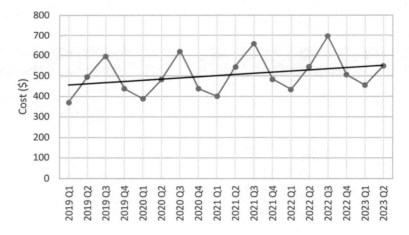

a Overall, the trend is increasing/decreasing.

b This data is weekly/monthly/quarterly/yearly.

c The cost of power is highest in quarter _____.

d The cost of power is lowest in quarter _____.

e Estimate the variation in the data: _____ − _____ = _____

f How much do you think power will cost this household in 2023 Q3? _____

2 This graph shows the mass of apples and kiwifruit exported from New Zealand between 1996 and 2022.

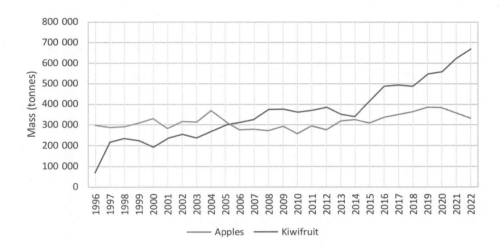

a This data is weekly/monthly/quarterly/yearly.

b Which fruit do you think we will export more of in 2024? _____

c Compare the trends for apples and kiwifruit.

_____

_____

3 The graph shows the number of marriages in New Zealand each quarter. Describe each feature of the data.

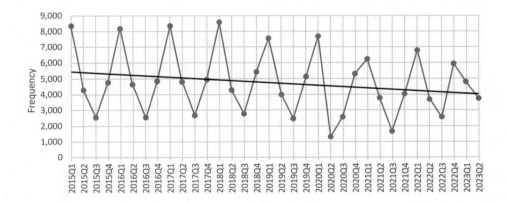

Pattern: _____

_____

Trend: _____

_____

Variation: _____

_____

4 The graph shows the number of tickets given for using a phone while driving, between the start of 2021 and the middle of 2023. Describe any patterns, trends and variation in the data.

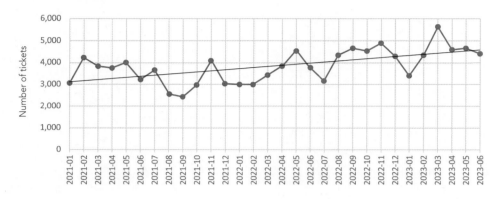

_____

_____

_____

_____

5 This graph shows the price per kg of kūmara, broccoli and avocado from 2012 until 2022.

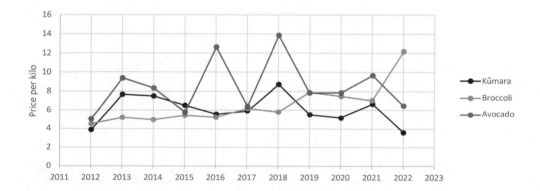

a This data is weekly/monthly/quarterly/yearly.

b Compare the prices of kūmara, broccoli and avocado.

_____

_____

_____

_____

_____

_____

_____

Scatter plots

- Scatter plots are used to display data for a **relationship** investigation.
- You need data for **two numeric variables** from **each** member of the population (or sample) and you analyse both at once to see if there is a relationship between them.
- A **scatter plot** has one variable on each axis and **each member** of the population or sample is represented by **one point**.
- The distribution of the points on the scatter plot is used to determine if there is a relationship (correlation) between the two variables.
- If a relationship exists, it can be used to make **predictions** or **estimates**.

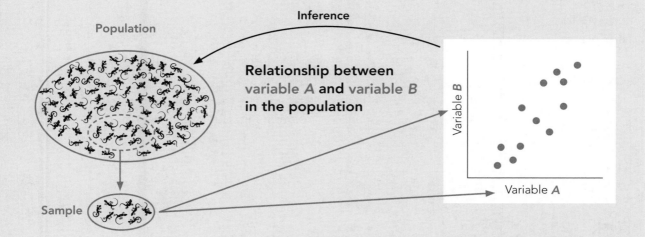

Features of scatter plots
Direction

Positive relationship:

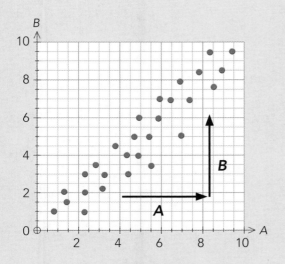

As A increases, B **increases**.

Negative relationship:

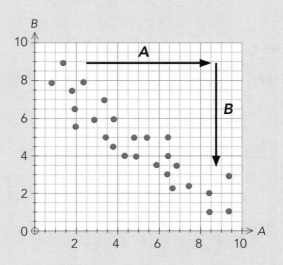

As A increases, B **decreases**.

Strength
The amount of scatter tells you the strength of the relationship.

Strong:

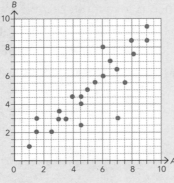

There is not a lot of scatter.
This means the relationship is strong.

Moderate:

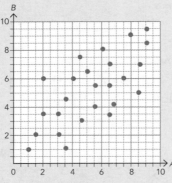

There is a moderate amount of scatter.
This means there is a moderate relationship but it isn't strong.

Weak:

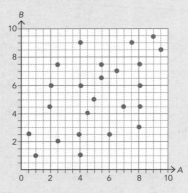

There a lot of scatter.
This means the relationship is weak.

Trend

Linear

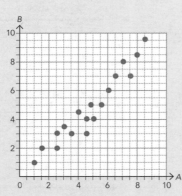

Non-linear

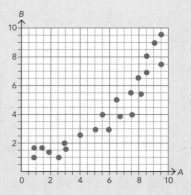

A line drawn through the middle of the data **would** be straight.
This means the data **is** changing at a constant rate.

A line drawn through the middle of the data **would not** be straight.
This means the data is **not** changing at a constant rate.

Example: The number of visitors and the daily maximum air temperature were recorded at a local pool over summer.

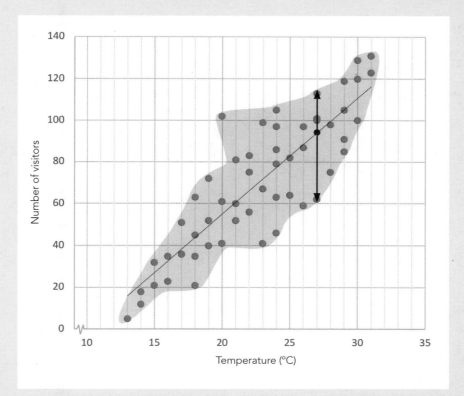

1 Describe the features of the scatter plot.

Summary sentence: There is a positive, strong and linear relationship between temperature and the number of tickets sold.

Direction: I notice that the relationship between the daily maximum temperature and the number of tickets sold is positive. This means as the temperature increases, the number of tickets sold also tends to increase.

Strength: I notice that there is not a lot of scatter between the points on the graph. This means there is a strong relationship between the temperature and number of tickets sold.

Trend: I notice the trend appears to be linear because I ruled a straight line through the middle of the data. This means that the relationship between temperature and tickets sold changes at a constant rate.

2 Estimate the number of tickets sold if the daily maximum air temperature was 27°C.

We would expect to sell about 94 tickets, but it could be anywhere between 60 and 110.

3 Describe any unusual features in the scatter plot.

Around 102 tickets were sold on a day that reached 20°C. This is unusual, as on most days of a similar temperature, only between 40 and 80 tickets were sold.

Answer the following questions.

1   The graph shows the relationship between left foot length and height for a group of high school students.

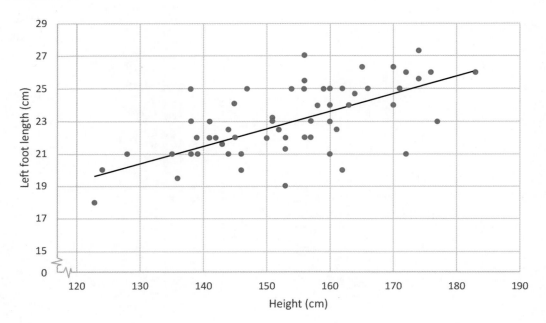

There is a _____, _____ and _____ relationship

between _____ and _____

Direction: _____

_____

_____

_____

Strength: _____

_____

_____

_____

Trend: _____

_____

_____

Prediction: _____

_____

_____

2   This graph shows the average number of chicks per breeding pair of little blue penguins in Oamaru and the date the eggs were laid.

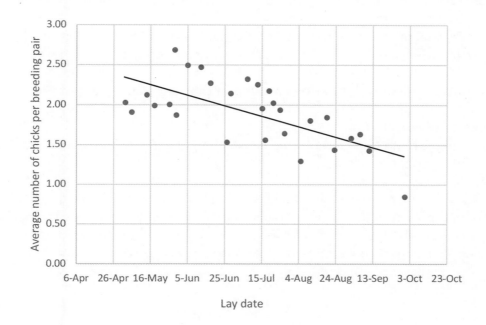

There is a _____, _____ and _____ relationship

between _____ and _____

Direction: _____

_____

_____

_____

Strength: _____

_____

_____

_____

Trend: _____

_____

_____

Prediction: _____

_____

_____

_____

Correlation and causality

- If two variables are **correlated**, change in one variable is associated with change in the other variable.
- If there is a **causal** relationship between two variables, the change in one variable **causes** a change in the other variable, e.g. how much you exercise affects your fitness.

That two variables are correlated does not mean that one causes the other.

Multiple causes of correlation

The media often attribute one factor as the cause of an effect, when in fact there are several contributing and often linked factors.

Example: Consider the headline 'No breakfast affects students' performance' (*Wall Street Journal*). There is no doubt that levels of blood sugar and hunger influence the performance of a student. However, whether a child has breakfast is also strongly associated with factors such as poverty, which also influences performance at school, as would absenteeism.

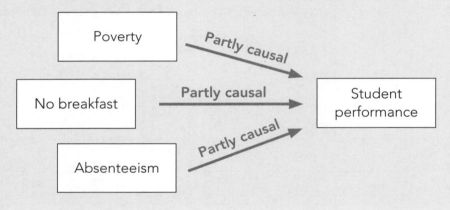

1 Suggest some other factors that might account for the relationships implied by these headlines.

Statement	Other factors
Drinking Coke leads to tooth decay.	Other dietary habits. Teeth-brushing habits.
Jumping on a trampoline causes joint issues.	
Smart people are healthier.	

Lurking variables
- A **lurking variable** is a variable that is associated with changes in **both** the original variables.

This graph shows the relationship between the number of TV sets owned per 1000 people and the average lifespan of people in a range of countries.

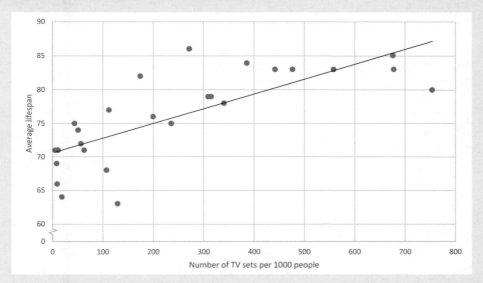

So, does high ownership of TV sets **cause** people in a country to live longer? If a country wants to increase the longevity of people, should it buy more TV sets?
High TV ownership indicates that a country is wealthy, so diets, medical services, living conditions, etc. are likely to be better. Therefore people in such countries live longer.

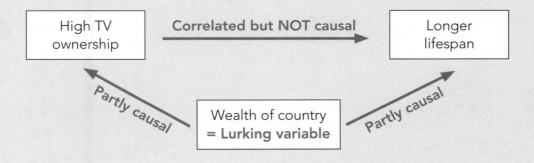

2   Suggest some lurking variables that might account for these relationships.

Relationship	Lurking variable
There is a positive relationship between the number of ice creams sold and the number of drownings.	
There is a positive relationship between the amount of popcorn consumed by cities and the number of serious crimes committed.	
There is a positive relationship between a person's height and their salary.	

Box plots

- **Comparing** data sets can enable us to make judgements and come to conclusions about **differences between groups**.
- Graphs that can be used for this include dot plots, box plots or back-to-back bar graphs.
- Which graph you choose will depend on the type of data that you have.
- Dot plots and box plots are commonly used for comparison data.

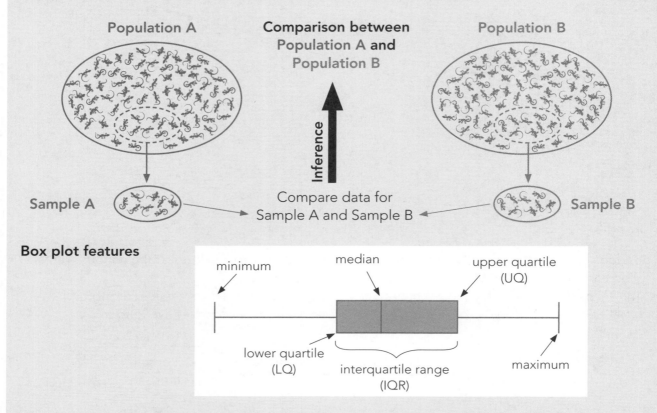

Box plot features

Comparing box plots
- We can compare box plots and make a suggestion about whether we are likely to see a difference in the population.
- This is possible by using the position of the medians and boxes.

Toby grew two varieties of tomato plants. He recorded the masses of each tomato he harvested.

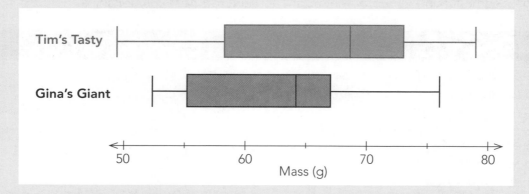

In this situation, we can make the call that Tim's Tasty tomatoes tend to be heavier than Gina's Giant tomatoes in the population. We can make this call, as half of the data for Tim's Tasty tomatoes is larger than three quarters of the data for Gina's Giant tomatoes.

   ISBN: 9780170477550

Answer the following questions.

1 The number of visitors to Auckland Zoo on Thursdays and Fridays is shown below.

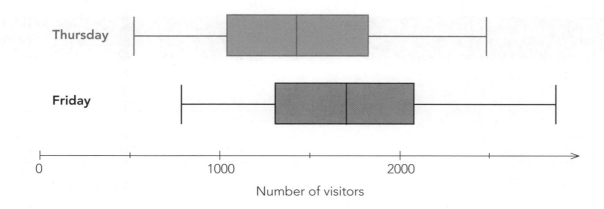

What conclusion can you come to about the number of Auckland Zoo visitors on a Thursday compared to a Friday?

_____

_____

_____

_____

2 This box plot shows the lengths of Hector's dolphins.

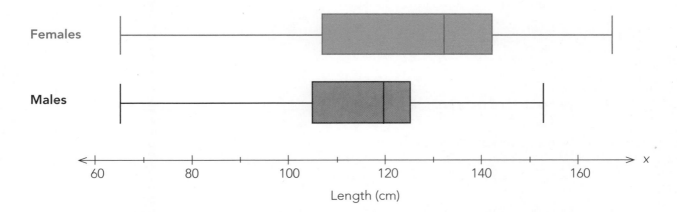

What conclusion can you come to about the length of female and male Hector's dolphins?

_____

_____

_____

_____

Which graphs should I use?

Tui collected this information from her classmates.

Year level	Height (cm)	Number of pets	Arm span (cm)	Shoe size	Favourite movie	Distance travelled to school (km)
11	143	4	151	6½	Titanic	0.9
12	182	0	176	10	Transformers	2.3
11	167	1	172	7	Avatar	1.8
12	158	3	161	8½	Ready Player One	4.6
11	164	2	160	8	Barbie	2.1
11	159	2	147	7	Jumanji	3.8
12	171	1	168	11½	The Hunger Games	0.5

Assuming you could collect more data from other class mates, answer the following questions.

1 **a** Name a variable from the table above that you could put in a bar graph.

_____

b What sort of graph would you use to display the heights of students?

_____

c Are there any variables that are suitable for a line graph? Explain your answer.

_____

_____

d What kind of data is shoe size? _____

e Name a pair of variables that would be appropriate for a scatter plot.

_____ and _____

f Describe how you could use comparative box plots to display some of this data.

_____

_____

g Are there any variables that might not be easily analysed? Why?

_____

_____

Infographics

- Infographics are modifications of traditional graphs or diagrams used to convey numerical information.
- They are often visually pleasing but they don't always make the message clear.

Answer the following questions.

1 The Ministry of Social Development created these graphics to show likely changes in the age structure of the New Zealand population between 2018 and 2034.

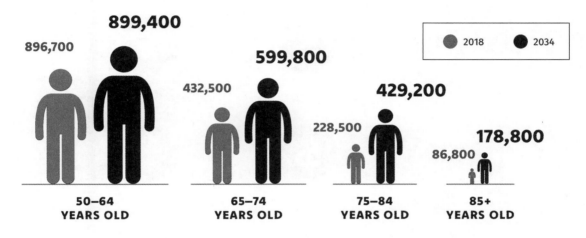

a What do you think the purpose of this infographic is? There may be several answers.

☐ To indicate that by 2034 our population will be taller.

☐ To indicate that the New Zealand population is aging.

☐ To suggest that the older you get, the smaller you become.

☐ To suggest that by 2034, people will be living longer.

☐ To indicate that the population of New Zealand will be greater in 2034.

b What 'traditional' type of graph could they have used instead?

☐ Bar graph   ☐ Scatter plot

☐ Box plot   ☐ Dot plot

c Do you think that this data is displayed fairly? Explain your answer.

_____

_____

_____

_____

2   The Ministry of Social Development also created this graphic to show the average net worth of
    New Zealanders of different ages.

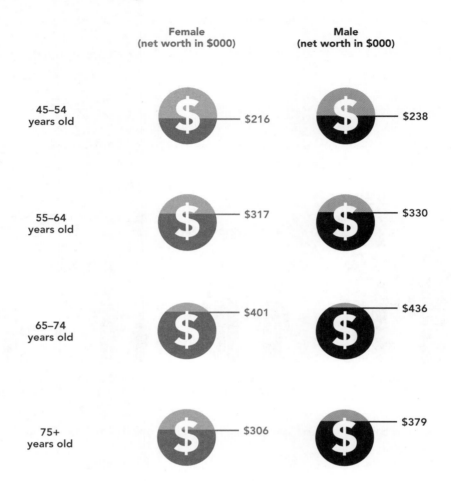

	Female (net worth in $000)	Male (net worth in $000)
45–54 years old	$216	$238
55–64 years old	$317	$330
65–74 years old	$401	$436
75+ years old	$306	$379

a   What average net worth do you think a completely shaded circle would represent? _____

b   What is the average net worth of 45–54 year old males? _____

c   Which age range has the highest average net worth? _____

d   Do males or females have a higher average net worth? _____

e   Name at least two 'traditional' graphs that could have been used to display this
    information.

    _____ or _____

f   Do you think that this infographic shows the data fairly and clearly? Explain your answer.

    _____

    _____

    _____

3 Auckland Transport (AT) compiled this information from the 2018 census on ways people travel to work or their education facility.

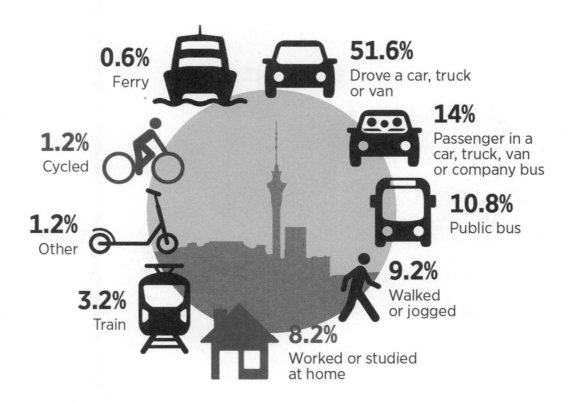

0.6% Ferry

51.6% Drove a car, truck or van

1.2% Cycled

14% Passenger in a car, truck, van or company bus

1.2% Other

10.8% Public bus

3.2% Train

9.2% Walked or jogged

8.2% Worked or studied at home

a What is the most popular way of getting to work or education facilities?

_____

b Name at least two 'traditional' graphs other than a pie graph that could have been used to display this information.

_____ or _____

c This information could have been displayed on a pie graph. Give one reason why a pie graph would have been a better way to show this data.

_____

d Give two drawbacks to showing this data on a pie graph.

i _____

ii _____

e Do you think that this infographic shows the data fairly and clearly? Explain your answer.

_____

_____

_____

4 This infographic showing the languages spoken by New Zealanders has been made using data from the 2018 census. It describes the population as if New Zealand had 100 people in total.

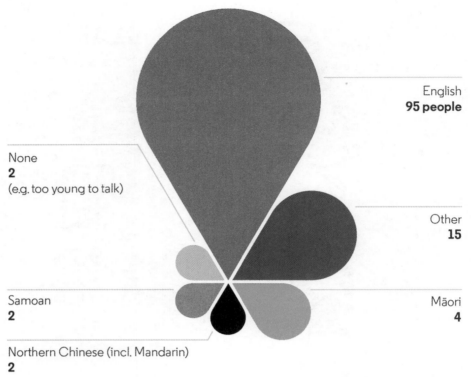

English
95 people

None
2
(e.g. too young to talk)

Other
15

Samoan
2

Māori
4

Northern Chinese (incl. Mandarin)
2

* Multiple answers possible so will total more than 100.

a What is the most commonly spoken language? _____

b In 2018, the population of New Zealand was 4.901 million. How many people spoke Māori in 2018?

_____

c Name a 'traditional' type of graph that they could have used instead.

_____

d Explain why a pie graph would not be appropriate for this data.

_____

e Do you think that this infographic shows the data fairly and clearly? Explain your answer.

_____

_____

_____

ISBN: 9780170477550

5 This infographic shows information about horses as companion animals in New Zealand during 2020.

Companion horses in New Zealand, 2020

There are over

70,000

companion horses in New Zealand

2%

of NZ households share their home with at least one horse

a In 2020, there were 1 856 000 households in New Zealand. How many households share their home with at least one horse?

_____

b The average cost of a horse was $3000, but a quarter of these were free. Calculate the average cost of horses that were not free.

_____

c How many horses were microchipped?

_____

d Do you think that this infographic shows the data fairly and clearly? Explain your answer.

_____

_____

_____

e Suggest a different way of showing the percentages, and explain why you think it would be better.

_____

_____

6 In 2022, the New Zealand Police had 15 000 members of the public surveyed. They were asked:
   a if they would support Police as a career option (No, Unsure or Yes), and
   b to rate their trust and confidence in the police on a scale of 1 to 5.

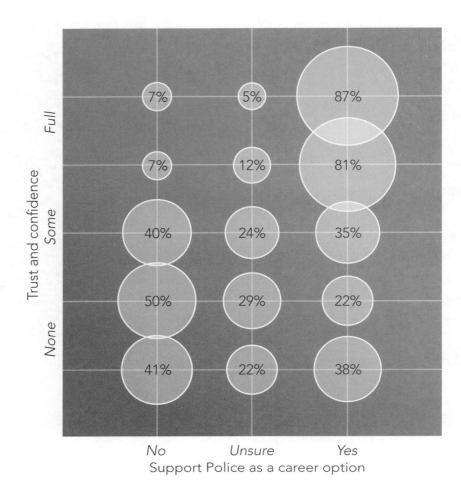

a Of the respondents who had some trust and confidence in the police (scored 3), what percentage would support Police as a career option?

_____

b Of the respondents who would support Police as a career option, what percentage had the lowest level of trust and confidence in the police? Explain why this question cannot be answered.

_____

_____

c 'Most people have trust and confidence in the police and would support Police as a career option.' Do you agree with this statement? Explain your answer.

_____

_____

_____

 Data analysis fundamentals

- There are two measures that we need to know in order to be able to discuss and compare distributions:

  1 Where is the **centre** of the data?
  2 How widely is the data **spread**?

Measures of centre

- There are **three** measures for the centre of the data.

Name	Calculation	Advantages	Disadvantages
Mean	$\dfrac{\text{the sum of all the data values}}{\text{the number of data values}}$	• Easy to calculate	• Distorted by values that are vastly different from the others.
Median	middle value	• Usually a good measure of centre	• Data must be put in order
Mode	value that occurs most frequently	• Very easy to find	• Very unreliable as measure of centre • Often there are two or none

Mean

- The numbers **do not need to be in order** for this calculation.
- Means are not always whole numbers, so **sensible rounding** may be needed.
- The mean is influenced by unusually large or small values.

$$\text{mean} = \frac{\textbf{sum of all the data values}}{\textbf{number of data values}}$$

Example:

Notice that 0 must be included in the calculation.

0 2 5 9 1 7 4 7 61

$$\text{mean} = \frac{0 + 2 + 5 + 9 + 1 + 7 + 4 + 7 + 61}{9}$$
$$= 10.\dot{6}$$

There are 9 numbers in the data set.

The mean of this data set is 10.$\dot{6}$ or 10.7 (1 dp)

Notice what happens to the mean if the 61 is removed.

0 2 5 9 1 7 4 7

$$\text{mean} = \frac{0 + 2 + 5 + 9 + 1 + 7 + 4 + 7}{8}$$
$$= 4.375$$

There are now 8 numbers in the data set.

The mean of this data set is 4.375

The 61 was so much larger than the other values that it influenced the mean significantly.

Median

- If there is an **odd number of values** in a data set, the median is the **middle number**.
- If there is an **even number of values**, the median is **halfway between the two middle numbers** in the data set.
- Before you can calculate the median, you must **put the data in order**.

Examples:
1 **A data set with an odd number of values**

$$7 \quad 13 \quad 9 \quad 14 \quad 21 \quad 18 \quad 24 \quad 11 \quad 19$$

Put them **in order** before finding the median: **7 9 11 13 14 18 19 21 24**

The median of this data set is 14

This is the middle number.

2 **A data set with an even number of values**

$$1 \quad 3 \quad 4 \quad 8 \quad 9 \quad 10 \quad 15 \quad 16 \quad 17 \quad 17 \quad 20 \quad 21$$

The median of this data set is $\dfrac{10 + 15}{2} = 12.5$

Mode

- The mode is the **most common value**.
- Sometimes there are **several modes**.
- If there are **three or more** numbers that occur equally often, we say there is **no mode**.

Examples:
1   (8) 17 20 9 3 (8) 14 12 7 9 (8) 17 2

The most common number is **8**: there are three of them.
The mode of this data set is 8

2   15 6 (12) 11 3 (9) 8 14 (12)(9) 19 17 4

Both **9** and **12** occur twice.
The modes are 9 and 12

3   (24) 28 (20) (25) 21 (20) 32 27 (25) (24) 29

There are three numbers that occur equally often: **20**, **24** and **25**.
If there are **three or more** modes, we say the data is **polymodal**.

 ISBN: 9780170477550

Answer the following questions.

1 Nina bought 5 pears for $4.10 at a market stall. At another stall she bought another 3 pears for $2.70. Calculate the mean cost of a pear.

_____

_____

2 Jed has two brothers. His brothers are 2 year younger and 4 years younger than he is. Jed is 12 years old. What is the median age of the brothers?

_____

_____

3 The mean temperature for the last three days has been 17°C. Give an example of what the temperatures could have been for the past three days.

_____

_____

4 Beatrice has a mean Science test score of 78% from two tests. What percentage would she need to get on the third test to increase her mean to 80%?

_____

_____

5 Lucy claims she is better at maths than her brother. Here are their scores for the last five tests.

Lucy	39	85	91	67	36
Her brother	57	72	71	51	67

Do you agree with Lucy?

_____

_____

6 The mean score for a class test was 57%. If every student had scored an extra 3%, the mean score would be:

☐ the same   ☐ > 57% and ≤ 60%   ☐ 60%

7 The mean age of a class of 28 students is 15. When the teacher's age is added, the mean of all their ages increases to 16. How old is their teacher?

☐ 28   ☐ 35   ☐ 40   ☐ 44

8 The mean of 50 observations was found to be 63. When the data was checked, it was found that a 49 was wrongly written as 94. The new corrected mean will be:

☐ 60.14   ☐ 61.1   ☐ 62.1   ☐ 63.9

9 The mean age of the boys in a class is 15.6. The mean age of the girls is 15.1. If the mean age for the whole class is 15.4, calculate the ratio of boys to girls in the class.

☐ 2:1   ☐ 3:2   ☐ 4:3   ☐ 5:4

Measures of spread

Range
- The range is the **maximum** value **minus** the **minimum** value in the data set.
- Note: the range is a **single number**.
- Like the mean, the range is affected by unusually large or small values.
- The data does not need to be in order to calculate the range.
- The range is a measure of the **variability** of the data.

$$\text{range} = \text{maximum} - \text{minimum}$$

Quartiles
- The upper quartile (UQ) is the **middle (median)** value of the **top half** of the data.
- The lower quartile (LQ) is the **middle (median)** value of the **bottom half** of the data.
- The interquartile range (IQR) is also a measure of the **variability** of the data.
  It is usually considered to be a better measure of spread than range because it isn't affected by extreme values.

$$\text{interquartile range (IQR)} = \text{upper quartile (UQ)} - \text{lower quartile (LQ)}$$

Examples:
1 If the number of pieces of data is **odd**, **exclude** the median from the top and bottom halves of the data.

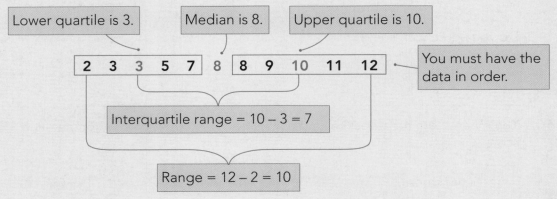

Lower quartile is 3.　　Median is 8.　　Upper quartile is 10.

2　3　3　5　7　8　8　9　10　11　12

You must have the data in order.

Interquartile range = 10 – 3 = 7

Range = 12 – 2 = 10

2 If the number of pieces of data is **even**, find the median values of the top and bottom halves of the data.

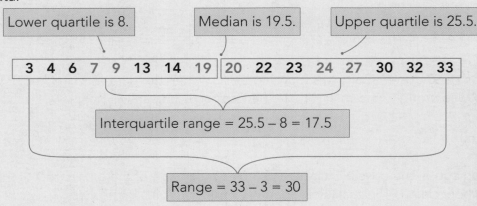

Lower quartile is 8.　　Median is 19.5.　　Upper quartile is 25.5.

3　4　6　7　9　13　14　19　20　22　23　24　27　30　32　33

In practice, whether you include or exclude the median values makes very little difference with a large data set.

Interquartile range = 25.5 – 8 = 17.5

Range = 33 – 3 = 30

Ranges and interquartile ranges are used to measure the variability of groups of data.

Answer the following questions.

1   7 9 14 18 19 25 27 29 33 37 40 47 49 49

Minimum = _____   LQ = _____   Median = _____   UQ = _____   Maximum = _____

Range = _____   Interquartile range = _____

2   If the single highest point were removed from a data set:

		Yes/Maybe/No
a	The mean will decrease.	
b	The median will decrease.	
c	The IQR will stay the same.	
d	The range will stay the same.	

3   Match the sets of data to the box plots.

a    _____

b _____

c _____

d _____

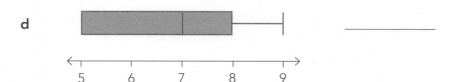

1	5, 5, 5, 5, 7, 7, 7, 8, 8, 8, 8, 9, 9	2	5, 5, 6, 6, 7, 7, 7, 7, 7, 7, 7, 8, 9
3	5, 5, 6, 6, 6, 7, 7, 8, 8, 9, 9, 9, 9	4	5, 5, 6, 6, 6, 6, 6, 7, 8, 8, 8, 9, 9

4   Indicate whether the following statements are true or false (✓ or ✗).

☐ The data set C has the largest IQR.   A

☐ Set B has a smaller range but larger IQR than set C.   B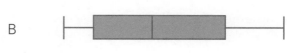

☐ Set A has a larger IQR and smaller range than set C.   C

Probability fundamentals

Probability
- Probabilities can be written as fractions, decimals or percentages.
- You can convert between these with your calculator.

The range of values for probabilities

0	0.5	1
Impossible	50:50 chance	Certain

Comparing probabilities
- When you want to compare probabilities, unless the denominators are the same, it is easiest to use decimals.

Example: Which probability is more likely: $\frac{2}{3}$ or $\frac{7}{11}$?

$$\frac{2}{3} = 0.\dot{6} = 0.666... \qquad \frac{7}{11} = 0.\dot{6}\dot{3} = 0.6363...$$

Two thirds is larger than seven elevenths, so it is slightly more likely.

Calculating probabilities

$$\text{probability} = \frac{\text{number of favourable outcomes}}{\text{total possible outcomes}}$$

Example: Students were asked to select a language they wanted to learn.

Te reo Māori	Japanese	Samoan	French	Spanish
23	9	15	11	16

What is the probability that a student wanted to learn te reo Māori?

> This is shorthand notation for 'probability a student chose te reo Māori'.

$$P(\text{Te reo Māori}) = \frac{23}{74} = 0.3108 \text{ (4 dp)}$$

Combining probabilities
- If you want to calculate the probability of one event **or** another, you **add** their probabilities, provided the two groups do not overlap.

Example: What is the probability a student wanted to learn te reo Māori or Samoan?

$$P(\text{te reo Māori or Samoan}) = \frac{23}{74} + \frac{15}{74} = \frac{38}{74} = 0.5135 \text{ (4 dp)}$$

Complementary events
- Complementary events occur when you are considering only **two** possible outcomes.
- The probabilities of complementary events always **add to 1**.

Example: If the probability that it will rain tomorrow is 0.35, then what is the probability that it will not rain?

$$P(\text{no rain}) = 1 - 0.35 = 0.65$$

Expected number

Predictions or estimations can be made based on probabilities.

$$\text{expected number} = P(\text{event}) \times \text{number of trials}$$

Examples:

1 Companion Animal New Zealand regularly undertakes surveys to collect data on the companion animal population. It estimates that there is a total of 851 000 dogs in the country.

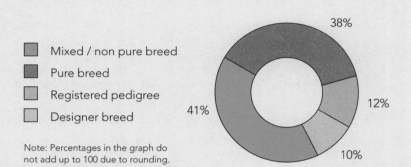

- Mixed / non pure breed
- Pure breed
- Registered pedigree
- Designer breed

Note: Percentages in the graph do not add up to 100 due to rounding.

a How many dogs in New Zealand would you expect to be a mixed/non pure breed?

41% of 851 000 = 0.41 x 851 000
= 348 910, so about 350 000.

b How many would you expect to be either designer or pure breed?

(10% + 38%) of 851 000 = 0.48 x 851 000
= 408 480, so about 408 000.

2 The probability of Eric being late to work is 0.04. In the next 80 days, how many times would you expect him to be late?

P(late) = 0.04 x 80 = 3.2
We would expect him to be late on 3 or 4 days.

> Rounding to a whole number is often sensible.

Answer the following questions.

1 Bridget bikes to school 45% of the time. How many times in the next 50 days would you expect her to bike to school?

_____

2 Around 0.6% of people are born with two different-coloured eyes (heterochromia). The population of New Zealand is around 5.3 million. How many New Zealanders would you expect to have heterochromia?

_____

3 Around one in 2000 children born each year have webbed toes (syndactyly). How many of the 58 887 babies born in 2022 would you have expected to have webbed toes?

_____

4 This graph is also from the Companion Animal survey. It estimates there are 1.2 million cats in New Zealand.

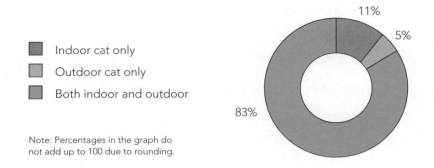

■ Indoor cat only

■ Outdoor cat only

■ Both indoor and outdoor

Note: Percentages in the graph do not add up to 100 due to rounding.

11%

5%

83%

a How many cats would you expect to be indoor only? _____

b How many cats would you expect to be outdoor? _____

5 In 2020, there were approximately 282 839 secondary school students in New Zealand. This table shows exclusion rates for different areas of the country.

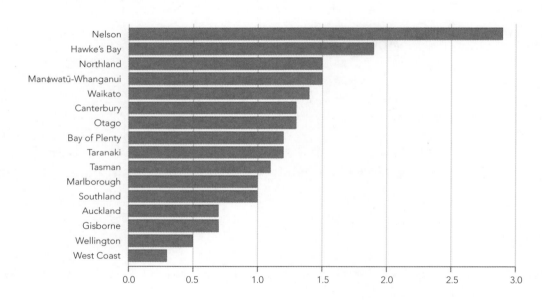

Exclusion rate for students in New Zealand schools

By region, 2020, age-standardised number of exclusions per 1,000 students

a If the rate of exclusions from Nelson were applied to the whole of New Zealand, how many students would you expect to be excluded?

_____

b The Nelson region is next to the Tasman region. Discuss their rates of exclusion.

_____

_____

Calculations from displays

- You can often take information from tables and graphs and use it for further calculations.
- You could be asked to find probabilities, percentages or expected numbers.

Example: Students were surveyed about their favourite subject at school.

a For what percentage of students was Science their favourite subject?

$\frac{26}{132} \times 100 = 19.7\%$ (1 dp)

b If a student was picked at random, what is the probability their favourite subject was English?

P(English) = $\frac{18}{132}$ = 0.1364 (4 dp)

c Another school has a roll of 784 students. How many of them would you expect to have Mathematics as their favourite subject?

$\frac{12}{132} \times 784 = 71.27$ (2 dp)

Therefore 71 or 72 students.

Subject	Number of students
English	18
Science	26
Mathematics	12
Social Studies	7
Health	15
Art	23
Physical Education	31
Total	**132**

d Comment on any assumptions that you have made in your previous answer.
I have assumed that the distribution of favourite subjects was the same in both schools. This is unlikely because favourite subjects often depend on the teachers, so are likely to differ.

Answer the following questions.

1 This graph shows babies born in New Zealand with the names Ayla, Indie and Violet.

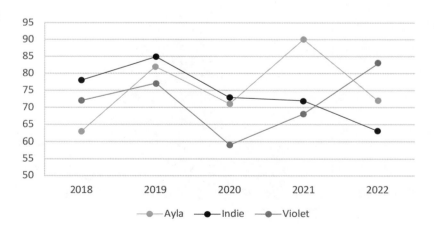

a Which was the most popular name in 2020? _____

b Which name has decreased in popularity over the years? _____

c There were 58 887 babies born in 2022. What percentage were named Ayla? _____

2 When someone has an accident in New Zealand, they can make a claim for support to ACC (Accident Compensation Corporation). This pie graph shows the locations where ACC injuries occurred in 2022.

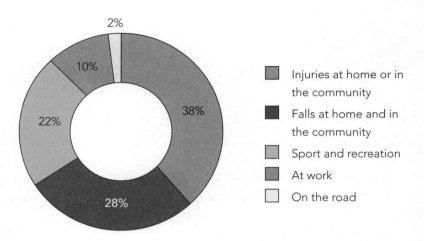

- ☐ Injuries at home or in the community
- ☐ Falls at home and in the community
- ☐ Sport and recreation
- ☐ At work
- ☐ On the road

a What were the most common locations for accidents in 2022? _____

b What percentage of accidents were at work or on the road? _____

c How else could this information have been displayed? _____

d In 2022, the total number of ACC claims was 1 878 862. How many of these would you expect to be the result of sport or recreation?

_____

e Tanya tripped over in her garden and broke her wrist. Explain a possible source of confusion when she has to tick the location for her accident.

_____

_____

3 This graph shows the times for the 71 competitors who finished the 2020 Tokyo Olympic women's marathon.

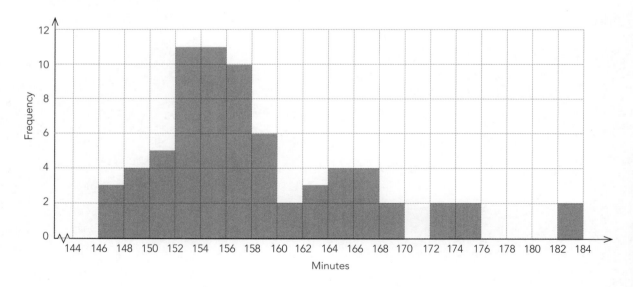

a In what interval was the winning time for the marathon? _____

b How many runners took 2 hours and 50 minutes or longer? _____

c What percentage of runners finished in less than two and a half hours? _____

4 The data shows the basic colours of the 5 669 853 cars that were registered in New Zealand in late 2022.

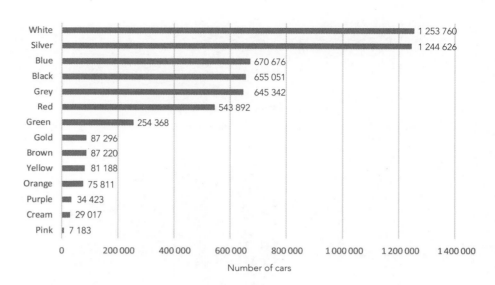

White 1 253 760
Silver 1 244 626
Blue 670 676
Black 655 051
Grey 645 342
Red 543 892
Green 254 368
Gold 87 296
Brown 87 220
Yellow 81 188
Orange 75 811
Purple 34 423
Cream 29 017
Pink 7 183

Number of cars

a What type of variable is this data? _____

b What percentage of registered cars in New Zealand were white in 2022? _____

c If a car is picked at random, what is the probability that it is pink or purple? _____

d If 2000 cars pass a point, how many would you expect to be green? _____

5 This graphic shows the number of female and male employees and their annual salaries at a government ministry.

a How many female employees are there?

_____

b What percentage of female employees earn over $120 000?

_____

c How does this compare with male employees?

_____

_____

d An employee is selected from the group earning $60 001 to $80 000. What is the probability that the employee is male?

_____

e How else could you display this data?

_____

_____

_____

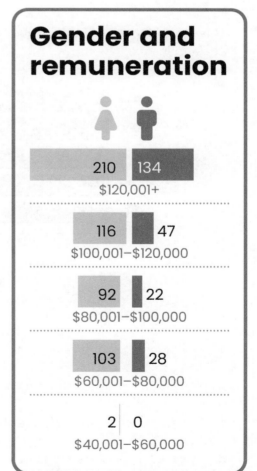

Gender and remuneration

210	134	
$120,001+		
116	47	
$100,001–$120,000		
92	22	
$80,001–$100,000		
103	28	
$60,001–$80,000		
2	0	
$40,001–$60,000		

6 The Ministry of Education collects information on students' education in New Zealand. The graph shows the number of homeschooled students of different ages in July 2022. The total number of students being homeschooled was 10 945.

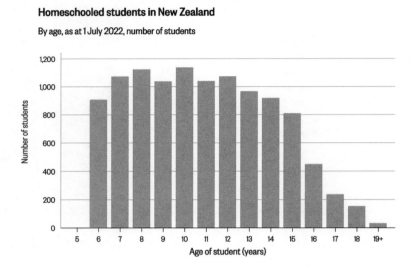

Homeschooled students in New Zealand

By age, as at 1 July 2022, number of students

a Approximately how many six-year-old students were homeschooled? _____

b Five hundred homeschooled students were randomly chosen. How many would you expect to be 17 years old or older?

_____

7 This graph shows the conservation status of some groups of native land animal species in New Zealand in 2022.

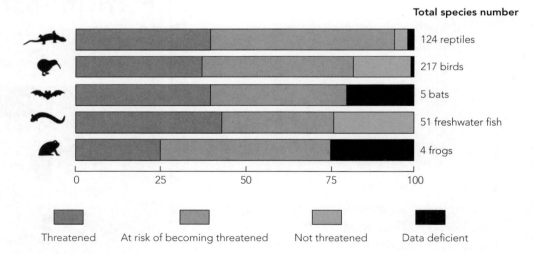

a For which group of animals is the lowest proportion threatened? _____

b How many species of bats are threatened? _____

c Estimate the number of reptile species that are either threatened or at risk. _____

d Which group has the largest **number** of threatened species? _____

 ISBN: 9780170477550

Two-way frequency tables

- When there is data for two different characteristics for each member of a population, you can display these in a two-way table.

Example: A survey of school students was taken about what size bed they have and if they have their own room or share with others:

	Double bed	Single bed
Own bedroom	12	34
Shared bedroom	3	21

It helps to add totals to your table:

	Double bed	Single bed	Totals
Own bedroom	12	34	46
Shared bedroom	3	21	24
Totals	15	55	70

> Check that both add to 70.

a What is the probability that a student has their own bedroom?

$$P(\text{own bedroom}) = \frac{46}{70} = 0.6571 \text{ (4 dp)}$$

b What is the probability that a student has a double bed in their own room?

$$P(\text{double bed in own room}) = \frac{12}{70} = 0.1714 \text{ (4 dp)}$$

c If two students from this school are selected at random, what is the probability that both share their bedrooms?

> One student **and** the other share a bedroom so you **multiply** their probabilities.

$$P(\textbf{both} \text{ students share a bedroom}) = \frac{24}{70} \times \frac{23}{69}$$

$$= \frac{4}{35} \text{ or } 0.1143$$

> Once a student has been selected, there are only 69 left.

d If a student **who shares** a room is selected at random, what is the probability that they also have a double bed?

$$P(\text{double bed in a shared room}) = \frac{3}{24} = 0.125$$

> We are concerned just with the 24 students who share a bedroom.

e At another school, with a roll of 849 students, how many would you expect to have a single bed in a shared bedroom?

$$\text{Expected number with single bed in shared room} = \frac{21}{70} \times 849$$

$$= 254.7$$

$$= 254 \text{ or } 255 \text{ students}$$

f What assumption have you made in **e**?
I have assumed that both schools are in areas that are economically similar.

Use the data in the following tables to answer the questions. Where needed, round your answers to 3 sf.

1 Tongue rolling is the ability to roll the edges of the tongue to make a tube-like shape. Some people are unable to do this. It is said to be a genetic trait. Complete the table.

	Year 11	Year 9	Totals
Can roll tongue		87	136
Can't roll tongue	16		
Totals		122	

a What percentage of students can roll their tongue? _____

b What is the probability that a randomly selected student was in Year 9 and can't roll their tongue?

_____

c What is the probability that a Year 11 student can roll their tongue?

_____

d If two students from Year 11 were selected at random, what is the probability that both can roll their tongue?

_____

2 Jackie surveyed 168 students. She asked them whether or not they had ever been surfing. Complete the table.

	Had been surfing	Had not been surfing	Totals
Year 13	21		88
Year 10		79	
Totals	34		

a What is the probability that a randomly selected student had not been surfing?

_____

b What is the probability that a randomly selected student was a Year 13 who had been surfing?

_____

c What is the probability that a randomly selected Year 10 student had not been surfing?

_____

d What is the probability that two randomly selected students from her survey had both been surfing?

_____

e The school has 257 Year 10 students. How many would you expect to have been surfing?

_____

3 The table below shows a survey of senior students and the type of driver's licence they have and whether or not they own a car.

	No licence	Learner	Restricted	Full	Totals
Car	3		28		
No car		35		8	217
Totals	70	51			341

a What is the probability that a student has no licence and no car?

_____

b What is the probability that a student with a restricted licence has no car?

_____

c What is the probability that a student who doesn't have a car has a restricted licence?

_____

d What is the probability that a student who has a car has a restricted or full licence?

_____

e What percentage of students have either a restricted or full licence?

_____

4 A survey was done on Year 11 students about their rules at home. Students identified if they had household jobs (chores) and whether or not they had a curfew.

	Chores	No chores	Totals
Curfew	218		
No curfew		103	
Totals	295		524

a What is the probability that a student had no curfew?

_____

b What is the probability that a student has curfew and chores?

_____

c What is the probability that a student who has no curfew, also has no chores?

_____

d What is the probability that two randomly selected students have chores and a curfew?

_____

e There are approximately 62 000 Year 11 students in New Zealand. How many of these would you expect to have neither chores nor a curfew?

_____

Question styles

- There are a number of styles of questions you may be asked. You need to be able to answer them appropriately.

Providing evidence when answering the question

- You should always try to provide **numerical** evidence to back up your answers.
- Other ways of asking for evidence: 'Justify your answer.'
  'Support your conclusions with ...'
- Focus on the **main** message from the graphics, not the detail.

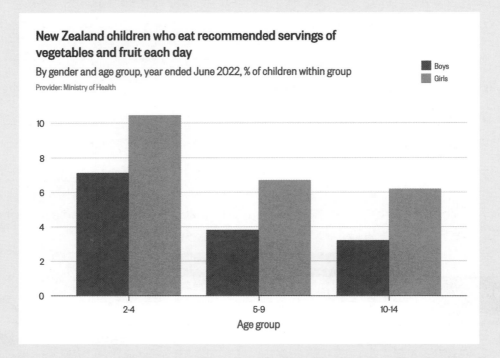

New Zealand children who eat recommended servings of vegetables and fruit each day

By gender and age group, year ended June 2022, % of children within group

Provider: Ministry of Health

Legend: Boys / Girls

Observations
- For all age groups of children more girls than boys eat the recommended servings of vegetables and fruit each day.
- The percentage of children who eat the recommended servings of vegetables and fruit each day drops between the ages of 2 and 14.

> There are two main messages from this graph.

Evidence
- For 2–4 year olds, just over **7%** of boys eat the recommended servings of vegetables and fruit each day compared with just over **10%** of girls.
- For 5–9 year olds, these figures have **dropped** to just under **4%** of boys and under **7%** of girls.
- They **dropped** again for 10–14 year olds: just over **3%** for boys and just over **6%** for girls.

> Notice that **percentages** are estimated for each group to support both observations.

> Notice the word '**dropped**' has been used to support the second observation.

Answer the following questions.

1   The graph shows the retail price of mandarins in New Zealand between January 2021 and June 2023.

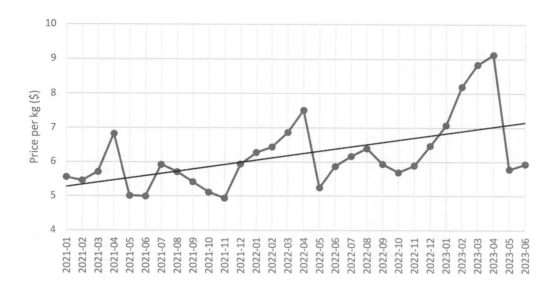

a   Which are the two main messages from this graph?

   1   The price of mandarins in December 2021 was $6 per kg.

   2   Generally, mandarins are most expensive in April and cheapest in May.

   3   Over this period, the average price of mandarins has increased by about $2 per kg.

   4   The data shows a consistent peak in April of each year with prices around 50c more than in March.

b   Describe the evidence for each of your chosen sentences.

   Sentence _____: _____

   _____

   _____

   _____

   Sentence _____: _____

   _____

   _____

   _____

2 The graph shows the percentage of New Zealanders with each type of driver's licence in each age range.

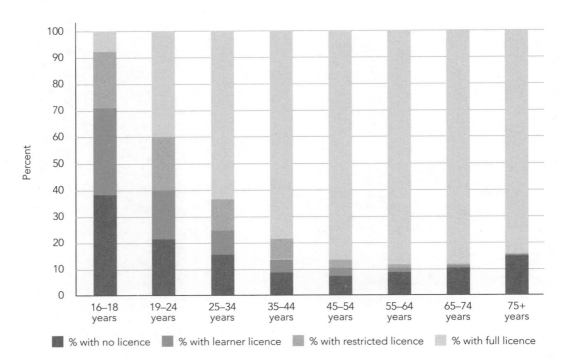

a Which are the two main messages from this graph?

i The proportion of people with a full licence increases with age up until 64 years, after which it doesn't change much until people reach 75 years. ☐

ii About 85% of 75+ year olds have a full driver's licence. ☐

iii The proportions of people with no licence decreases with age until 54 years, after which it increases. ☐

iv 75+ year olds tend to lose their full driver's licence. ☐

b Describe the evidence for each of your chosen sentences.

Sentence _____: _____

_____

_____

_____

Sentence _____: _____

_____

_____

_____

3 The graph shows the percentage of New Zealand emissions that come from each sector of the economy.

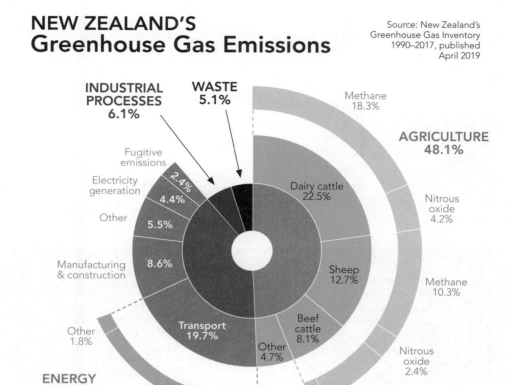

NEW ZEALAND'S Greenhouse Gas Emissions

Source: New Zealand's Greenhouse Gas Inventory 1990–2017, published April 2019

Note: Percentages in the graph may not add up to 100 due to rounding.

Write down two significant messages that this graphic is intending to tell the reader, and use evidence from the graphic to justify each.

i _____

_____

_____

_____

_____

ii _____

_____

_____

_____

_____

4 These graphs show the population of Wellington by age and sex in 1867 and 2013.

Note: The age breakdown varies for the two periods.

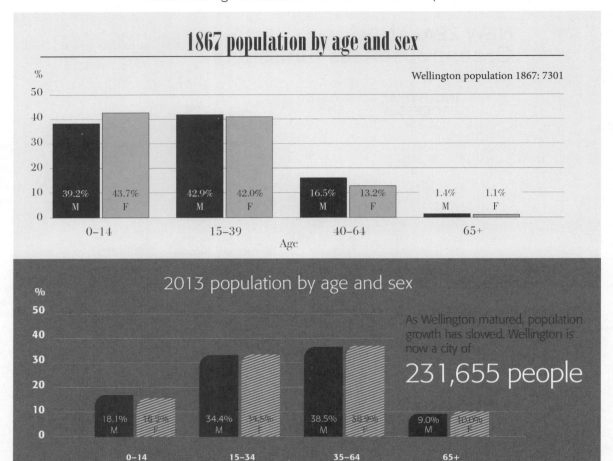

1867 population by age and sex

Wellington population 1867: 7301

Age	0–14	15–39	40–64	65+
M	39.2%	42.9%	16.5%	1.4%
F	43.7%	42.0%	13.2%	1.1%

2013 population by age and sex

As Wellington matured, population growth has slowed. Wellington is now a city of

231,655 people

Age	0–14	15–34	35–64	65+
M	18.1%	34.4%	38.5%	9.0%
F	18.3%	34.5%	38.9%	10.0%

Write down two significant points that can be concluded from these graphics, and use evidence from them to justify each. Hint: Check the x axes carefully as the scales may affect your answers.

i _____

_____

_____

_____

_____

ii _____

_____

_____

_____

_____

Give a number of supporting statements

- You could be asked to provide **more than one** piece of evidence or supporting statement.
- Make sure you provide **at least** the number asked for.
- It helps the marker if you clearly bullet point or number your separate pieces of evidence or statements.

This graph shows the differences between the average temperature each year and the average for all the years between 1961 and 1990.

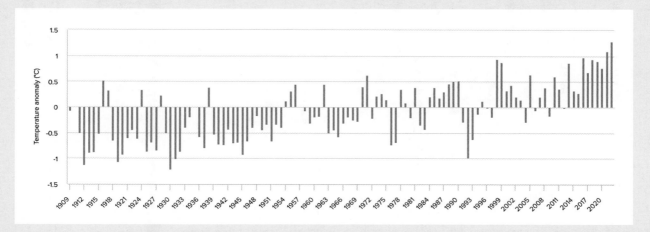

'The New Zealand climate is becoming warmer.'
Give at least **three** different supporting statements.

> If you are asked for a specific number of statements, make sure you give **at least** that number.

- Between 1909 and 1960, there were only eight years (just one year in six) when the temperature was higher than average, but between 1996 and 2022, temperatures have been higher than average in 21 out of 26 years.

> You need to provide evidence, so use **numbers** from the graphic to support your statements.

- Between 1909 and 1960, only one year had an average temperature that was more than 0.5° above average, but since 1996 there have been 12 such years.

- Up until 1960, most years were colder than average, but since 1996 only five years (about one year in every five) have been colder than average.

> You could number or bullet point them, but make sure there are clear different statements given.

Answer the following questions.

1 This graph shows the price change in men's footwear each quarter between the start of 2016 and the middle of 2023.

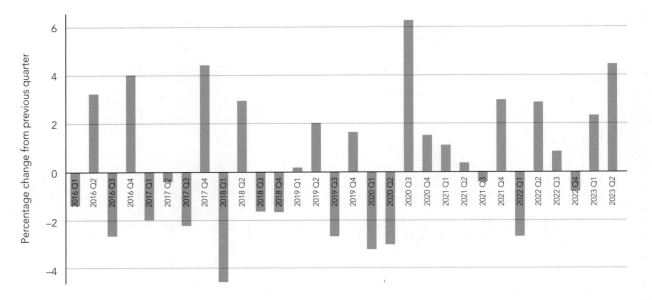

a 'It's cheapest to buy men's shoes in quarter 1 in New Zealand.'
Give at least two supporting statements.

i _____

_____

_____

ii _____

_____

_____

b 'Quarter 4 is the most expensive time of year to buy men's shoes in New Zealand.'
Give at least two supporting statements.

i _____

_____

_____

ii _____

_____

_____

2 This graph displays the number of seatbelt offences in the North and South Island of New Zealand.

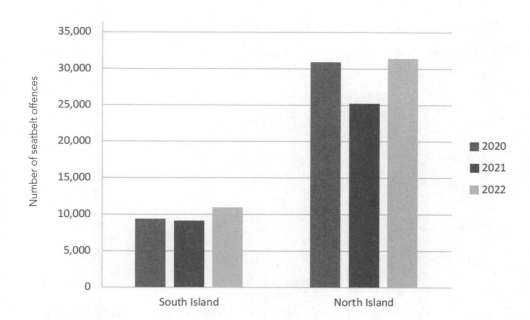

Gloria says that South Islanders are better at following road rules than the North Islanders. Decide whether you agree with this claim and give two supporting statements for your answer.

Decision: _____

i _____

_____

_____

_____

_____

_____

ii _____

_____

_____

_____

_____

3 This graph shows the number of males and females in New Zealand holding each driving licence type in 2018.

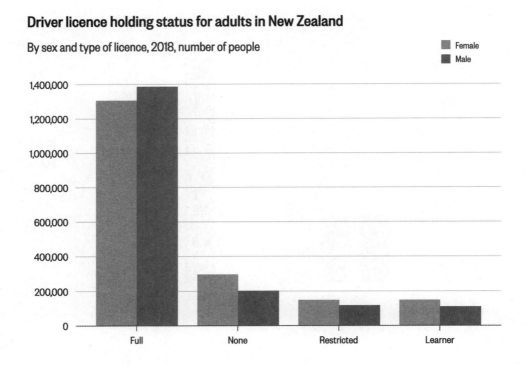

'Men are more likely to have a driver's licence in New Zealand.'
Decide whether you agree with this claim and give two supporting statements for your answer.

Decision: _____

i _____

_____

_____

_____

_____

ii _____

_____

_____

_____

_____

Agree, disagree or can't tell for sure

- When interpreting graphs, you need to explain your reasoning.
- You could be asked whether you agree or disagree with a statement, or whether you can't tell for sure.
- In some cases, any of these could be correct depending on your reasoning.
- Another way of asking this is: 'Comment on [a claim].'

Example:

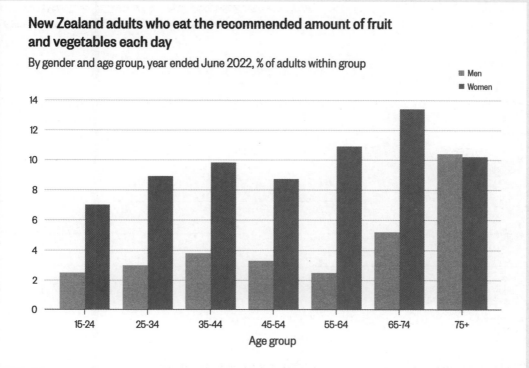

'Women eat more fruit than men.'

☑ Agree   ☐ Disagree   ☐ Can't tell for sure

Explain your answer.
There is a much higher percentage of women who eat the recommended amount of fruit and vegetables in all but one year group, so it is very likely that women (75+) eat more fruit than men.

Or:

☐ Agree   ☐ Disagree   ☑ Can't tell for sure

Explain your answer.
More women eat the recommended amount of fruit and vegetables. However, this graph doesn't show how much fruit is being eaten, and it doesn't distinguish between fruit and vegetables, so we cannot tell for sure.

Answer the following questions.

1 The graph shows the number of births registered with the most popular female names in New Zealand between 1848 and 2018.

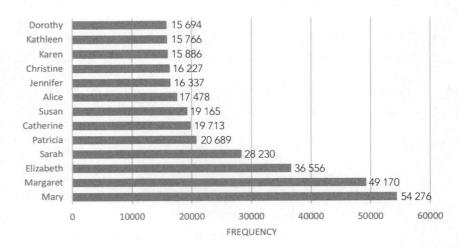

'"Karen" was the eleventh most popular name for baby girls during the 20th century.'

[ ] Agree   [ ] Disagree   [ ] Can't tell for sure

Explain your answer. _____

_____

_____

2 Consumer NZ did an investigation into washing machines.

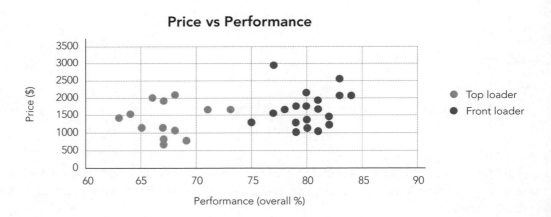

'There is no relationship between performance and price of washing machines.'
Using this graph as a basis, comment on this claim.

_____

_____

_____

_____

3 Coral says that vegetables are the most expensive they have ever been.

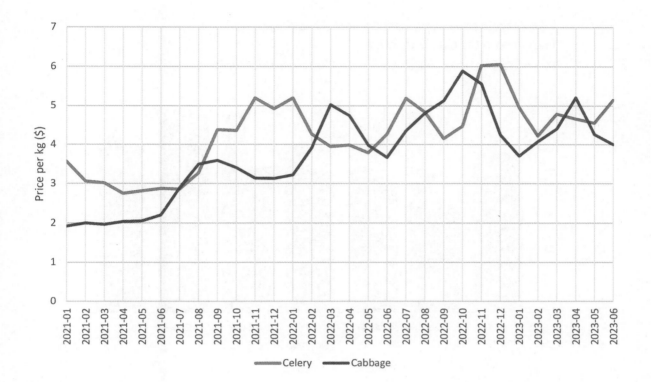

Give at least one reason why you could agree with or disagree with Coral's statement.

Agree

_____

_____

_____

_____

_____

_____

Disagree

_____

_____

_____

_____

_____

_____

4 'Younger people are less likely to visit a dentist.'

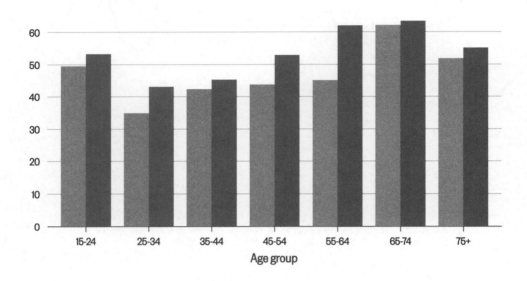

New Zealand adults who visited a dental health care worker at least once in the past 12 months

By gender and age group, year ended June 2022, % of adults with natural teeth

■ Men
■ Women·

Age group

Give at least one reason why you could agree with or disagree with this statement.

Agree

_____

_____

_____

_____

_____

_____

Disagree

_____

_____

_____

_____

_____

_____

Comparing values

- Statements could be made comparing two values or probabilities.
- It is common to see statements like 'twice as likely' or 'three times as likely'.
- These types of prompts are sometimes combined with claims.

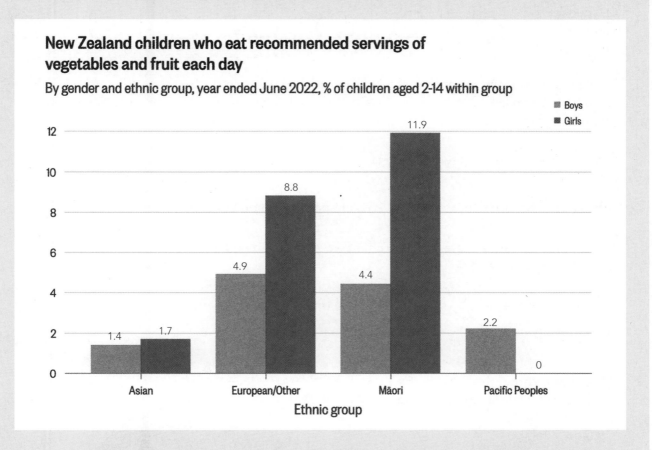

New Zealand children who eat recommended servings of vegetables and fruit each day

By gender and ethnic group, year ended June 2022, % of children aged 2-14 within group

'Māori girls are more than twice as likely as European boys to eat the recommended servings of fruit and vegetables each day.'

Only 4.9% of European boys meet the requirement whereas 11.9% of Māori girls do. This means that the claim can be supported because $\frac{11.9}{4.9}$ = 2.43 (2 dp).

Māori girls are more than twice as likely to meet the recommended servings of fruit and vegetables as European boys.

Do **not** just write an answer. You must show your working as well.

'European girls are three times as likely to eat the recommended servings of fruit and vegetables each day as Asian girls.'

8.8% of European girls meet the requirement whereas 1.7% of Asian girls do.

European girls are actually more than five times as likely ($\frac{8.8}{1.7}$ = 5.18) to meet the recommended servings of fruit and vegetables as Asian girls.

Answer the following questions.

1 Here is the weather forecast for the next week in Pleasantville.

	Mon	Tues	Wed	Thurs	Fri	Sat	Sun
Chance of rainfall	10%	60%	25%	20%	10%	15%	5%

Huia made the statement 'It is at least twice as likely to rain on Tuesday as it is on Wednesday.' You have been asked to agree or disagree with Huia and why. Which of these responses would be the best?

a I agree because it's more likely to rain on Tuesday than not rain. ☐

b I agree because there is only a 25% chance of rain on Wednesday. ☐

c I agree because there is a 60% chance of rain on Tuesday and only 25% on Wednesday. ☐

d I agree because $\frac{60}{25}$ is 2.4 so it is more than twice as likely to rain on Tuesday as it is on Wednesday on probabilities. ☐

2 The graph shows the level of the highest educational qualification attained by people living in New Zealand.

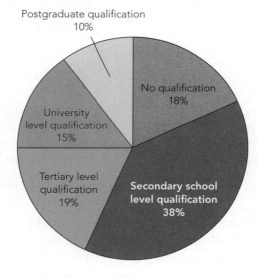

'New Zealanders are twice as likely to leave school with a qualification as with no qualification.' Do you agree with this statement?

_____

_____

_____

_____

_____

3   The table shows the results of an NZTA Waka Kotahi study on the driver's licence pass rates in various locations (2015–2018).

Places with the <u>highest</u> pass rates	Places with the <u>lowest</u> pass rates
Blenheim 81%	Manukau 48%
Invercargill 79%	Westgate 53%
Hāwera 78%	New Lynn 55%
Taupō 78%	Lichfield St 56%
Nelson 76%	Frankton 57%

Gregory said he was twice as likely to pass his driver's licence test in Blenheim than if he were to sit it in Manukau. Do you agree with this statement? Justify your answer.

_____

_____

_____

_____

_____

4   The graph shows the average house prices in some regions in 2022.

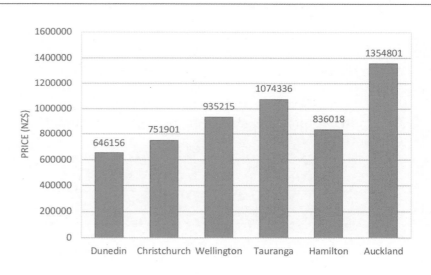

A newspaper headline reads 'Houses in Dunedin half the price of Auckland houses'. Is this justified?

_____

_____

_____

_____

_____

5   Environmental Science & Research (ESR) collects data on notifiable diseases in New Zealand. The following table shows the incidence of some diseases which are typically contracted from food and water contamination during the start of 2023.

Disease	Jan	Feb	Mar	Apr	May	Jun	Jul	Aug	Sep	Totals
Campylobacteriosis	763	484	480	328	366	442	424	520	456	**4263**
Cryptosporidiosis	42	36	36	21	54	44	38	94	261	**626**
Gastroenteritis	20	30	57	31	41	38	26	41	38	**322**
Giardiasis	86	80	95	64	100	71	53	86	56	**691**
Salmonellosis	85	95	96	47	73	70	46	71	58	**641**

Comment on the following claims using the information in the table.

a   'During 2023, you were twice as likely to get Campylobacteriosis in January as in any other month.'

b   'In September 2023, the number of Cryptosporidiosis cases was unusually high.'

c   'Campylobacteriosis was the most dangerous disease contracted from food and water contamination during 2023.'

Similarities and differences

- You could be asked to compare graphs or situations.
- You need to describe both what is **similar** and what is **different**.

Example: This graph shows the number of homeschooled students in New Zealand in 2021 and 2022 by age.

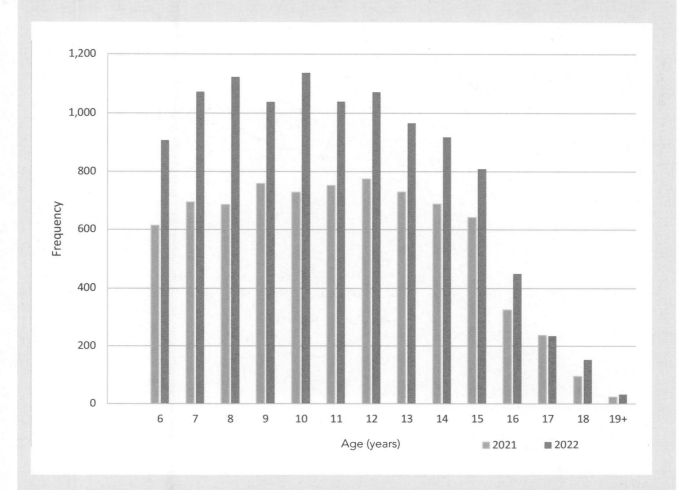

Similarities

- Both years follow a similar overall pattern. The numbers increase between six and seven years old and then remain reasonably steady until 12 years. After that, they drop by about 90 students per year between 12 and 15 years, then very steeply until there are very few by the age of 19 or older.

Differences

- The number of primary students (six to 12 year olds) homeschooled in 2022 was between 300 and 600 more than in 2021. During the first three secondary school years (13 to 15 years), there were about 200 more students in 2022 than in 2021. The only age group that dropped slightly in 2022 was 17 year olds.

Discuss the similarities and differences in the following data sets.

1 This graph shows the number of little blue penguins that are breeding at two different sites in Oamaru.

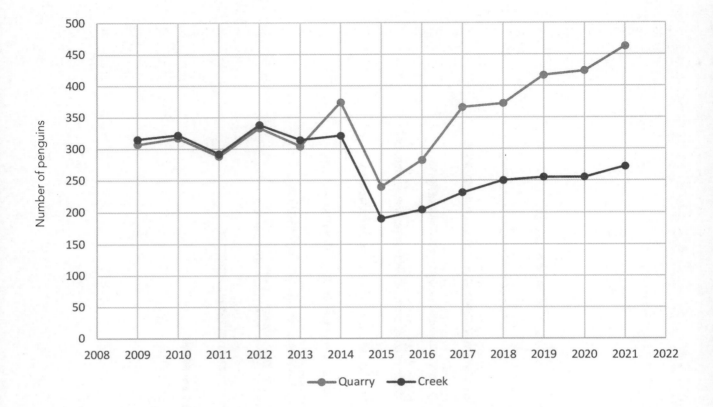

Similarities

_____

_____

_____

_____

_____

_____

Differences

_____

_____

_____

_____

_____

_____

  ISBN: 9780170477550

2 This graph shows the annual average retail prices of fuel per litre in New Zealand from July 2006 until July 2023.

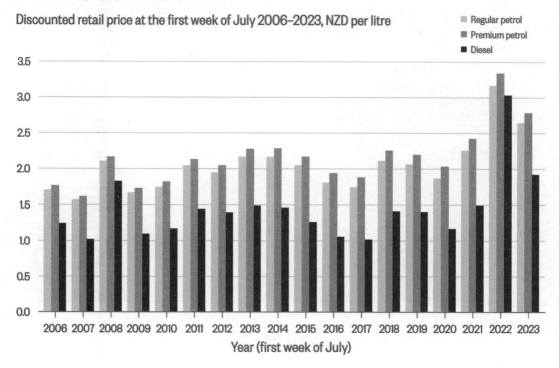

Retail fuel prices in New Zealand

Discounted retail price at the first week of July 2006–2023, NZD per litre

Legend: Regular petrol, Premium petrol, Diesel

Similarities

_____

_____

_____

_____

_____

_____

Differences

_____

_____

_____

_____

_____

_____

3 These graphs show the age composition of the populations of people living in the Thames-Coromandel district and in Hamilton city.

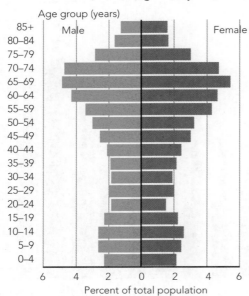

Thames-Coromandel district population pyramid
At 30 June 2019
(Median age = 54 years)

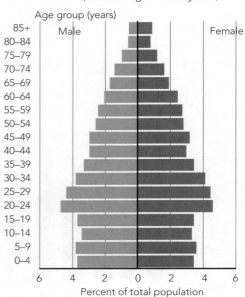

Hamilton city population pyramid
At 30 June 2019
(Median age = 32.3 years)

Similarities

_____

_____

_____

_____

_____

_____

Differences

_____

_____

_____

_____

_____

_____

 ISBN: 9780170477550

Assumptions and limitations

- When answering questions, you should discuss any **assumptions** you have made about the collection of data.

  Examples: That the sample size is adequate.

  That the selection of subjects is representative of the populations.

  That the data is fairly and accurately presented.

- You should also discuss any **limitations** to the use and applicability of the data.

  Examples: Was the sample big enough?

  Did the data come from a particular group or area which means conclusions can't be generally applied?

Example: This graph shows the reasons for leaving the last job for people who are currently unemployed in New Zealand.

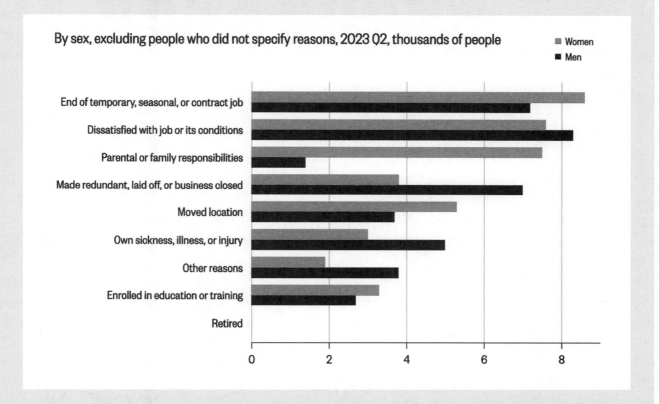

Assumptions
- People were asked for the **main** reason for leaving their job. Often there will be several reasons.

Limitations
- The data was collected for quarter 2 of the year. We cannot assume that the same pattern would occur at other times of the year, particularly for seasonal work.
- Some people did not specify a reason, so there is not data for every person who left a job and became unemployed.
- Some people listed 'Other reasons', so there is not data for every person who left a job and became unemployed.

Give at least one assumption and one limitation for each of the following sets of data.

1   This graph shows the average absence for holidays taken by school students during term two in New Zealand between 2015 and 2023.

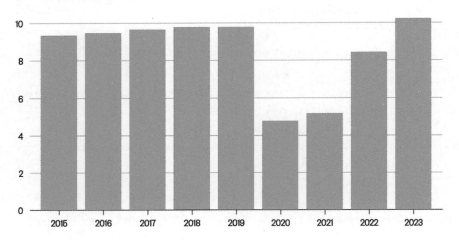

Term 2, 2015–2023, half-days

a   Assumption: _____

_____

b   Limitation: _____

_____

2   This graph shows the percentage of people reporting fever and cough symptoms. The data comes from an online self-reporting respiratory illness surveillance system in New Zealand.

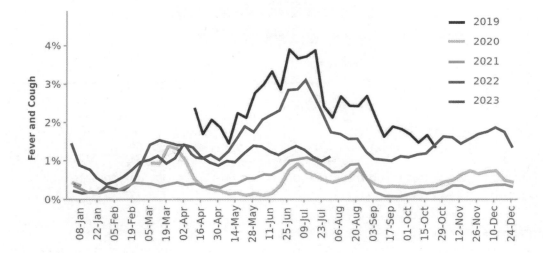

a   Assumption: _____

_____

b   Limitation: _____

_____

3 This graph shows the proportion of five-year-old children without cavities from three ethnic groups in the Taranaki DHB.

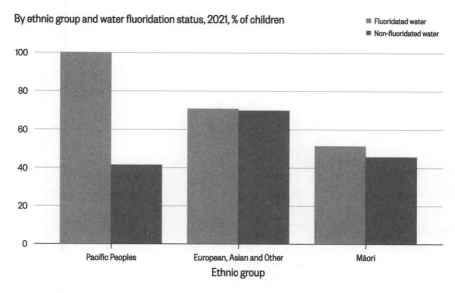

Give at least one assumption and one limitation for this graph.

_____

_____

_____

_____

_____

4 ESR collects data on notifiable diseases in New Zealand. Some of these are in the table on page 82. This is the process of how the data is collected:

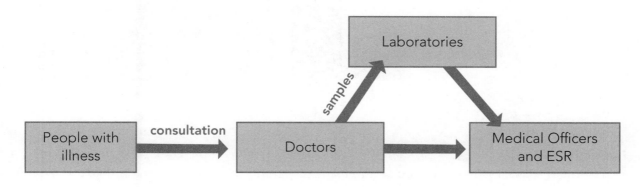

Give at least one assumption and one limitation of this process.

_____

_____

_____

_____

_____

I notice, I wonder

I notice …
- Discuss the **main features** rather than just stating information or numbers off the graph.
- Some things you could look for:
  — Maximum and minima
  — Similarities and differences
  — Unusual features
  — Patterns.

I wonder …
- Where did the data come from?
- Is the sample size adequate?
- How and where was the data collected? (Sample method)
- Who collected the data? Did they have a particular interest in the results?
- Why was it collected?
- What question might they have asked?
- Who was surveyed? How were they selected?
- Is the display suitable? Does it display the data fairly?
- How widely can we apply the findings/predictions?
- Who might this information be useful for?

Example: The graph shows the number of New Zealand men and women in six age groups who belong to occupational unions.

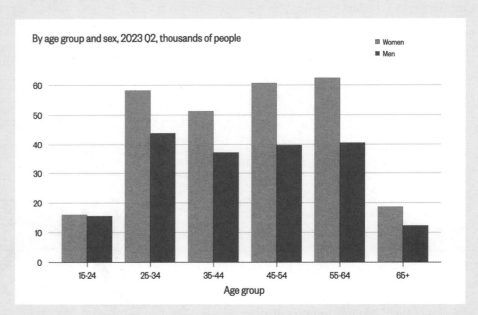

By age group and sex, 2023 Q2, thousands of people

I notice:
- That at all ages, women are much more likely to belong to a union for their main job in New Zealand.
- Very few 15–24 year olds have joined unions.
- Fewer people over the age of 65 belong to unions. This may be because many people retire at 65.

I wonder:
- Do all jobs have unions?
- Are women more likely to be in professions that have unions, e.g. teaching, nursing?
- Why are there fewer people under the age of 24 in unions? Are they more likely to be in jobs that have no unions?

 ISBN: 9780170477550

What do you notice and wonder about these graphs?

1. The Ministry of Social Development created this graphic. The figure to the left of the arrows is for 2018, and that to the right is the projected figure for 2034.

Our population is becoming more diverse

FOR THOSE AGED 65+

637,500 → **928,200**
European or other (including New Zealander)

59,500 → **171,900**
Asian

48,500 → **109,400**
Māori

21,300 → **46,700**
Pacific Peoples

4,000 → **18,000**
Middle Eastern / Latin American / African

2018   2034

I notice:

The majority of those aged 65+ are _____.

The number of Middle Eastern/Latin American/African people who are over 65 years will double/triple/more than quadruple.

There will be over _____ more New Zealanders who are over 65 by 2034.

The number of Europeans over 65 years increases by only 50%, whereas all the other groups at least double/triple/quadruple in numbers.

The number of Pacific people over the age of 65 will be less/more than double.

I wonder:

_____

_____

_____

_____

_____

_____

2 This graph shows the driving test passing rates on Saturdays by time of day in New Zealand. These figures covered all Vehicle Testing New Zealand stations for the period January 2015 to September 2018.

I notice:

_____

_____

_____

_____

_____

_____

I wonder:

_____

_____

_____

_____

_____

_____

3 The graph shows the composition (%) of the total value of cash ($) in circulation in New Zealand that is made up of each type of coin and note between 2000 and 2022.

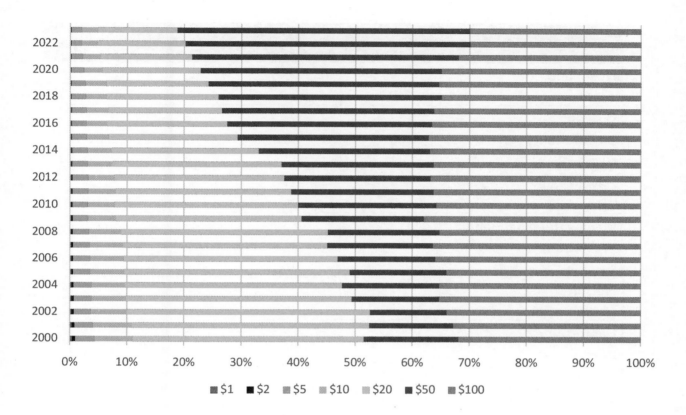

I notice:

_____

_____

_____

_____

_____

_____

I wonder:

_____

_____

_____

_____

_____

_____

4 When someone has an accident, they are often supported by the Accident Compensation Corporation (ACC). This table shows the number of ACC claims for washing line related injuries in each region between 1 January 2017 and 30 November 2022.

Region	2020	2021	2022	Population (x 100 000)
Northland	40	45	43	200
Auckland	218	202	193	1700
Waikato	99	93	76	500
Bay of Plenty	61	65	69	350
Gisborne	11	11	14	50
Hawke's Bay	46	31	30	180
Taranaki	27	25	19	120
Manawatū/Whanganui	37	46	40	260
Wellington	73	52	47	500
Tasman	6	9	7	60
Nelson	17	9	4	55
Marlborough	8	7	9	50
Canterbury	102	82	88	650
West Coast	–	7	–	30
Otago	48	49	38	250
Southland	15	11	8	100

I notice:

_____

_____

_____

_____

_____

_____

I wonder:

_____

_____

_____

_____

_____

_____

Misleading graphs

- Sometimes, published graphs are misleading or incorrect.
- This may be intentional or careless. Either way, it pays to think carefully about the information and who presented it.

Some common tricks:

1 Making graphs 3D

a 3D pie graphs

Students were asked what pet they would like to own.

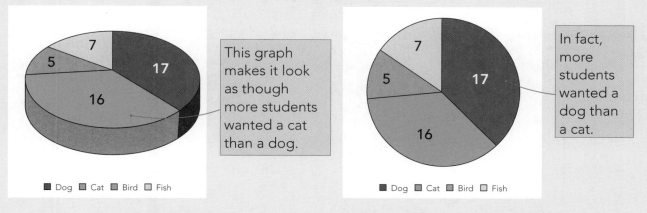

This graph makes it look as though more students wanted a cat than a dog.

In fact, more students wanted a dog than a cat.

These are misleading because the **front sector always looks bigger** than it should.

b 3D bar graphs that use perspective

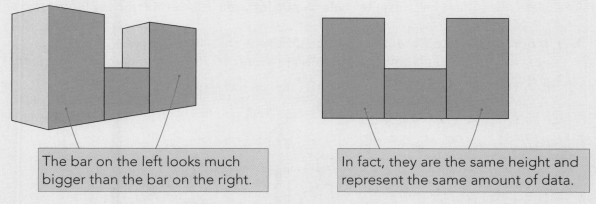

The bar on the left looks much bigger than the bar on the right.

In fact, they are the same height and represent the same amount of data.

These are misleading because they **make the bars look bigger or smaller, depending on which is 'closer'**.

2 Too much data on one graph

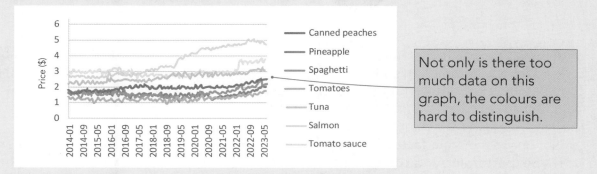

Not only is there too much data on this graph, the colours are hard to distinguish.

3 a Pictographs with scaling of 2D images

Pia recorded the number of bikes through two different intersections. She found that there were three times as many going through intersection B compared with intersection A.

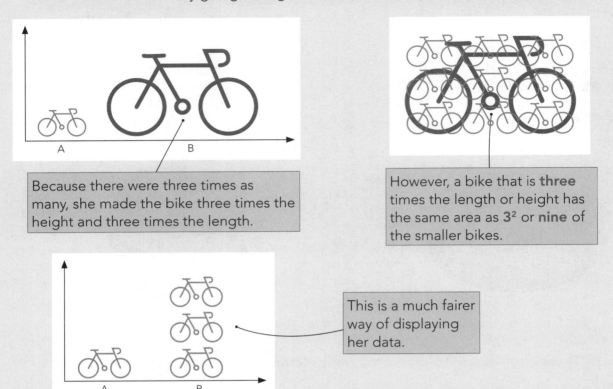

Because there were three times as many, she made the bike three times the height and three times the length.

However, a bike that is **three** times the length or height has the same area as 3^2 or **nine** of the smaller bikes.

This is a much fairer way of displaying her data.

These are misleading because the **scale factor for area = (scale factor for length)^2**.

b Pictographs with different-shaped or different-coloured images

This graph uses the heights of sports equipment to represent the number of students signed up to play.

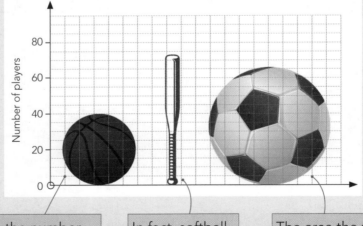

It looks as though the number of students wanting to play basketball is similar to softball, but it's closer to half.

In fact, softball is the sport most students want to play.

The area the football occupies makes it look as though many more students want to play this compared with softball.

Not only is the **scale factor for area** a factor here, but the visual impact can depend on **shape**, **patterning** and **colour**.

4 a Graphs with axes that don't start at zero

Results of vote for head student:

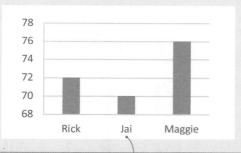 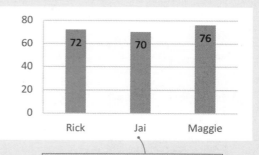

This graph makes it looks as though very few people voted for Jai, compared with the other two students. This is because the y-axis starts at 68 rather than 0.

In fact, there is not a lot of difference between the votes for head student.

These are misleading because the **relative heights of points or bars are changed.** **This is _very_ common.**

Note: Your axis doesn't have to start at zero if there is only one type of data and nothing is being compared.

b Graphs with inappropriate scales

- The scale of the axis should not exaggerate or diminish change.

These graphs show the annual mean amount of sea ice in Antarctica.

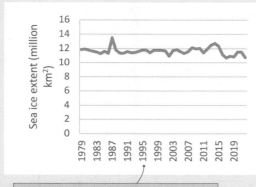 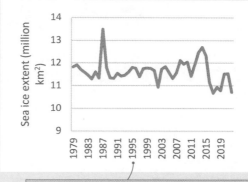

This graph makes it look like the sea ice isn't varying much.

This graph shows the same data with a different scale. This could lead to exaggeration.

5 Graphs that don't add up

People were asked to draw the apple logo from memory.

Very close to original		18%
Forgot to take the 'bite' out		14%
Drew bite on the wrong side		20%
Forgot the leaf		23%
Added a stalk		29%

These add to 104%. This indicates either errors or drawings were put in more than one category.

Describe what is misleading about these graphs.

1 Number of New Zealand children (2–14) eating fast food at least once a week.

_____

_____

_____

_____

_____

_____

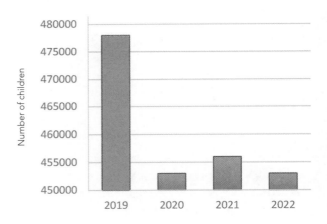

2 Selena wrote down the numbers of each type of animal that she saw at the zoo.

Animal	Frequency
Elephant	
Tiger	
Rhino	

_____

_____

_____

3 The graph shows the items bought at a cafeteria.

_____

_____

_____

_____

_____

_____

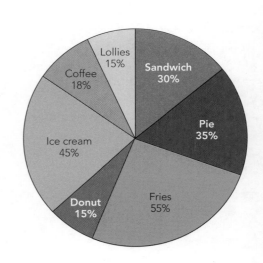

4 The graph shows the price of a packet of chewing gum in New Zealand between 2013 and 2023.

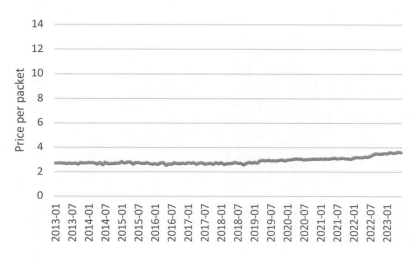

_____

_____

_____

5 This graph shows the number of people receiving benefits in New Zealand between 1996 and 2007.

_____

_____

_____

_____

_____

_____

_____

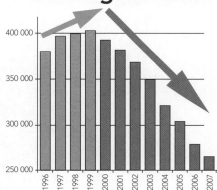

6 Students analysed the colours of cars in the staff park.

_____

_____

_____

_____

_____

_____

_____

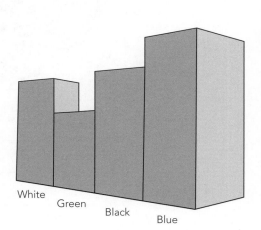

7 This graph shows the average height of females in a range of countries.

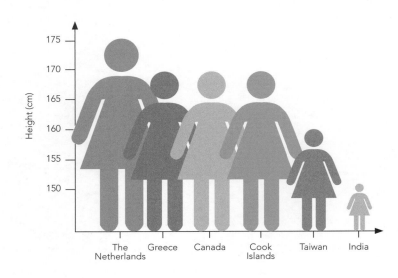

_____

_____

_____

_____

8 This graph shows the number of votes cast for each political party in the 2023 New Zealand election.

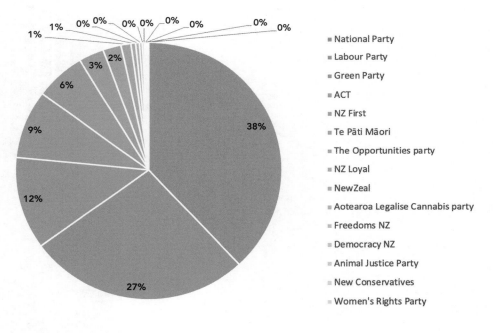

_____

_____

_____

_____

9 This graph shows what students prefer to drink.

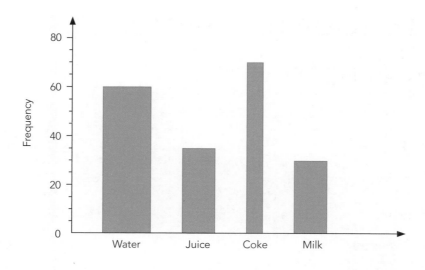

_____
_____
_____
_____

10 This graph shows the annual value ($NZ) of New Zealand's trade with Japan.

_____
_____
_____
_____

Number fundamentals

Percentages and GST

Percentages

Finding a percentage of a quantity

Example: Find 7% of 580.

> Remember, 'of' means **multiply**.

You could convert the percentage to a decimal:

$$7\% \text{ of } 580 = 0.07 \times 580$$
$$= 40.6$$

Or use the % function on your calculator:

$$7\% \times 580 = 40.6$$

> Not all calculators are the same, so you will need to experiment until you find how yours works.

Increasing by a percentage

Example: Increase 75 by 16%.

> This means that we need 100% plus 16%, or 116%.

> You could use your % button on your calculator.

Increased amount = 75 + (16% of 75)
$$= 75 + 12$$
$$= 87$$

or

Increased amount = 75 × 116%
$$= 75 \times 1.16$$
$$= 87$$

Decreasing by a percentage

Example: Decrease 240 by 28%.

> This means that we need 100% minus 28%, or 72%.

Decreased amount = 240 – (28% of 240)
$$= 240 - 67.2$$
$$= 172.8$$

or

Decreased amount = 240 × 72%
$$= 240 \times 0.72$$
$$= 172.8$$

Calculating a percentage

Example: A recipe for pasta contains 228 g eggs and 400 g flour. What percentage of the pasta is eggs?

$$\text{Percentage} = \frac{228}{628} \times 100$$
$$= 36.3\%$$

ISBN: 9780170477550

GST

- **GST** stands for **G**oods and **S**ervices **T**ax.
- It is added to everything you buy and it goes to the government to fund the running of the country.
- The standard current GST rate on all products and services is **15%**.

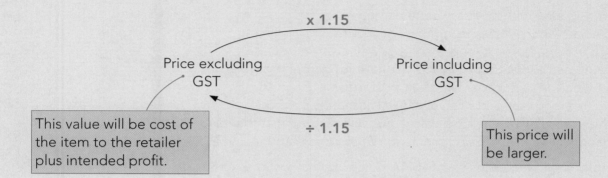

x 1.15

Price excluding GST → Price including GST

This value will be cost of the item to the retailer plus intended profit.

÷ 1.15

This price will be larger.

Examples:

1 Finding a GST inclusive price

The pre-GST price of a laptop is $4303.48. Calculate the retail price of the laptop.

$$\text{GST inclusive price} = \$4303.48 \times 1.15$$
$$= \$4949.00$$

This is the price including GST.

2 Finding the amount of GST paid on an item

A couch costs $1799 including GST. How much GST was paid?

$$\text{GST exclusive price} = \$1799 \div 1.15$$
$$= \$1564.35 \ (2 \ dp)$$
$$\therefore \ \text{GST} = \$1799 - \$1564.35$$
$$= \$234.65$$

This is the price without GST.

3 Finding the GST exclusive price

The amount of GST paid for a pair of headphones was $70. Calculate their GST exclusive price.

$$\text{GST exclusive price} \times \frac{15}{100} = \$70$$
$$\text{GST exclusive price} = \$70 \times \frac{100}{15}$$
$$= \$466.67 \ (2 \ dp)$$

This is the price without GST.

Rates and proportions

Direct proportions

- A rate is a relationship between two quantities.
- Rates are often written using the word '**per**'.
- '**Per**' means 'for each' and is another way of saying **divide**.
- Examples: the amount ($) paid per hour for gardening;
  the cost per kilogram of oranges;
  the speed of a vehicle in kilometres per hour.
- When we use two rates in a calculation, we assume that the relationship between the quantities in each remains constant. We assume that they are **in proportion**.

Example: 3 kg of oranges cost $10.47. Calculate the cost of 2 kg of oranges.

Method 1: Find the value of one unit and multiply this by the number of units you need.

3 kg cost **$10.47**, so 1 kg costs $10.47 ÷ **3** = **$3.49**

> One unit (1 kg) costs $3.49.

2 kg cost **2** x $3.49 = **$6.98**

Method 2: Use fractions.

3 kg cost **$10.47**
  copy ↓
2 kg cost **$10.47** x $\frac{2}{3}$ = **$6.98**

> Because you are trying to find a price, put the **price** on the **right**.

> Multiply by either $\frac{2}{3}$ or $\frac{3}{2}$. Use $\frac{2}{3}$ here because 2 kg will cost **less**.

Answer the following questions.

1 If 5 kg of apples cost $23.95, how much would 4 kg of apples cost?

_____

_____

2 If 1.5 kg of chicken drumsticks costs $7.47, calculate the cost of 2.3 kg.

_____

_____

3 Tia takes 24 minutes to walk 2 km. If she walks at the same pace, how long will it take her to walk 7 km?

_____

_____

4 It took five builders 7.5 hours to build a fence. How long would it have taken if there were only three builders working at the same rate?

_____

_____

5 Approximately 580 species of animals are being driven to extinction each week. How many are likely to be driven to extinction during December (31 days)? Round your answer appropriately.

_____

_____

6 An average of 96 elephants are killed for their ivory in Africa each day. How many are likely to be killed while you are at school today (7 hours)?

_____

_____

7 Beatrice did 43 press-ups in 2 minutes. If she can continue at this rate, how many press-ups could she do in 5 minutes?

_____

_____

8 Luciano can type 648 words in 9 minutes. How many words could he type in 7 minutes?

_____

_____

9 A baker can decorate 39 cupcakes in 25 minutes. How long will it take her to decorate 95 cupcakes?

_____

_____

10 Amy read 40 pages of a novel in an hour. At this rate, how long would you expect her to take to read a 300-page novel? Give your answer in hours and minutes.

_____

_____

Show your reasoning for these calculations.

11 Milk comes in various quantities. Which of the following is better value for money?

$6.29

1.5 L

$1.35

300 mL

_____

_____

12 On average, a person sheds 18.13 kg of skin during a 70-year life. Eric is 15 years old. How much skin would you expect him to have shed so far?

_____

_____

13 Three cricket players faced different numbers of balls. Jacinta faced 33 balls and scored 42 runs, Jude faced 98 balls and scored 124 runs, and Oliver faced 73 balls and scored 92 runs. Which player has the best run rate?

_____

_____

_____

Inverse proportions

- In the previous section, as one quantity got bigger, the other also got bigger, e.g. as the mass of oranges increased, the cost also increased.
- In some relationships, as one quantity gets bigger, the other **becomes smaller**, e.g. the faster you travel to school, the shorter the time it takes you.

Example: Olive drives home from work. She can travel at 46 kph and it takes her 24 minutes. How long would it take her to bike home from work at 21 kph?

Method 1:   Find the value of one unit and multiply this by the number of units you need.

At **46** kph she takes **24 min**, so at **1** kph it would take her **46** x 24 = **1104** min

At **21** kph she will take 1104 ÷ **21** = **52.6 min**

> Notice that you need to **multiply** because at 1 kph it will take her **longer**.

Method 2:   Use fractions.

At **46** kph she takes **24 min**

copy ↓

At **21** kph she takes 24 x $\frac{46}{21}$ = **52.6 min**

> Because you are trying to find a time, put the **time** on the **right**.

> Multiply by either $\frac{21}{46}$ or $\frac{46}{21}$. Use $\frac{46}{21}$ here because it will take her **longer** to walk.

Answer the following questions.

1   It took two trucks 38 minutes to pour a concrete pad. If they had three similar trucks, how long would it take to pour the pad?

2   A school camp has enough food for 124 people for seven days. How long will the food last if another 70 students join them?

3   Julie skated home from school at 7 kph and it took her 19 min. The next day she walked at 4 kph. How long did it take her?

4   It took a team of two tilers $4\frac{1}{2}$ days to tile a courtyard. How long should it take a team of five tilers to do the same job?

5   It took seven people 34 min to put desks out in the hall for exams. If there were only three people doing the same job, how long (in hours and minutes) would it take them?

6   A bath takes $12\frac{1}{2}$ minutes to fill at a rate of 10 L per minute. How many minutes would it take to fill at a rate of 15 L per minute?

Assumptions and limitations

- When answering questions, we should discuss any **assumptions** we have made: we often assume certain conditions apply.
- We also need to discuss any **limitations** to our answers: when or to whom would our answer not apply.

Example: It takes three housekeepers 4.5 hours to clean the hotel rooms. How long would it take two housekeepers to do the same job?

$$\frac{3}{2} \times 4.5 = 6.75$$

It would take two housekeepers six hours and 45 minutes to clean the rooms.

Assumptions:
- We assume that all the housekeepers are equally efficient.
- We assume that the same number of rooms were cleaned on each day and in similar locations, e.g. all on the same floor, not spread throughout a high-rise.

Limitations:
- This would only apply to this one hotel where rooms required similar degrees of cleaning.

Often, assumptions and limitations can overlap, so don't worry too much about distinguishing between them.

Answer the following questions.

1 **a** A recipe requires 500 g flour to make 16 scones. How many scones can be made with 850 g flour?

_____

b Discuss any assumptions and limitations of your answer.

_____

_____

_____

2 **a** A rock climber scaled a 12 m climb in 9.5 minutes. How long would it take him to climb 19 m?

_____

b Discuss any assumptions and limitations of your answer.

_____

_____

_____

3 **a** Sami knitted a 1.4 m scarf in 11.9 hours. How long would it take her to knit a 1.7 m scarf?

_____

b Discuss any assumptions and limitations of your answer.

_____

_____

_____

4 **a** Dennis plants five baby tomato plants and they produced a total of 63 tomatoes. How many tomatoes would you expect to get from six baby tomato plants?

_____

b Discuss any assumptions and limitations of your answer.

_____

_____

_____

5 **a** There is an average of five mice per hectare in forest areas. How many mice could you expect on an 80.5 ha island?

_____

b Discuss any assumptions and limitations of your answer.

_____

_____

_____

6 **a** Karearea (New Zealand falcon) have been recorded diving at speeds of 200 kph. How long would it take a karearea to dive 100 m (the length of a rugby pitch) to catch a rabbit?

_____

b Discuss any assumptions and limitations of your answer.

_____

_____

_____

Ratios

- Ratios show how an amount is split into several shares, often different sizes.
- Both parts of a ratio must use the **same units**.
- Ratios generally use whole numbers, not decimals or fractions, and are written in their simplest form.
- A **colon** (:) is used to separate the two numbers, e.g. 2:3 means 'two parts to three parts'.

Using ratios where the total is given

Example: $336 needs to be shared between two people in the ratio 3:5.

When there is a quantity that needs to be split using a ratio, follow these steps.

Step 1:	Find the total number of parts by adding the numbers in each share:	$3 + 5 = 8$
Step 2:	Find the value of one part by dividing the total by the number of parts:	$336 \div 8 = \$42$
Step 3:	Multiply the value of one part by each part of the ratio:	$3 \times \$42 : 5 \times \$42$
	So the ratio of money is	$\$126 : \$210$

So one person gets $126 and the other gets $210.

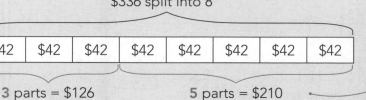

$336 split into 8

| $42 | $42 | $42 | $42 | $42 | $42 | $42 | $42 |

3 parts = $126   **5** parts = $210

Another way to think of this is $\frac{3}{8}$ of $336 is $126 and $\frac{5}{8}$ of $336 is $210.

Note: While the ratio doesn't have units or decimals, your answer could have these.

Share the quantities in the given ratios.

1 $756 in the ratio 3:4.

_____

_____

2 153 kg in the ratio 4:5.

_____

_____

3 $262.50 needs to be split between Lex and Max in the ratio of 2:3. How much should each person receive?

_____

_____

4 The ratio of colour-blind men to non-colour-blind men is 1:12. Out of 1524 men, how many would you expect to be colour-blind?

_____

_____

5 In a school with a roll of 980 students, 105 are left handed. What is the ratio of left-handed to right-handed students?

_____

_____

6 Todd, Tahlia and Kiwa got paid $690 for some painting. Todd worked for 2 hours, Tahlia for 1 hour and Kiwa for 3 hours. How much money should each of them receive?

_____

_____

7 In a box of 168 apples, two in every seven apples are rotten. How many rotten apples are there?

_____

_____

8 Keyana paid $2.00 towards a packet of 18 biscuits, and Mia paid the remaining $1.60. If the biscuits are shared fairly between them, how many should each get?

_____

_____

9 Three families combined to buy a 12 kg bag of potatoes. One family paid $10.50, the second paid $3.50, and a third paid $7.00. If the potatoes are divided fairly, what mass of potatoes should each family get? State any limitations to your answer.

_____

_____

10 a To make green paint, you mix blue paint and yellow paint in the ratio of 3:1. If you have 7 litres of blue paint and 4 litres of yellow paint, what is the maximum amount of green paint you can make?

_____

b How much blue and yellow paint will you have left over?

_____

Ratio calculations where one part is given

Examples:

1 Finding a part

Two sisters were given skateboards. The ratio lengths of the one sister's skateboard to the other's skateboard was 2:3. The smaller skateboard is 71 cm long. What length is the longer skateboard?

71 cm

Step 1:	Calculate the value of each part.	71 ÷ 2 = 35.5
Step 2:	Multiply the value of each part (35.5) by the number of parts (3).	35.5 x 3 = 106.5

So the longer skateboard is 106.5 cm long.

2 Finding a total

Alexa and Noa pooled their money to buy a bag of lollies. Alexa had $4 and Noa had $2. If they shared them fairly, and Noa got 19 lollies, how many lollies were in the bag?

Step 1:	Find the total number of parts by adding the numbers in each share (4 and 2).	4 + 2 = 6
Step 2:	Calculate the value of each part by dividing the number given (19) by the number of parts it represents (2).	19 ÷ 2 = 9.5
Step 3:	Multiply the value of each part (9.5) by the total number of parts (6):	9.5 x 6 = 57

So there were 57 lollies in the bag.

Answer the following questions.

1   Lucy and Angie delivered pamphlets in the ratio 4:3. If Lucy delivered 492 pamphlets, how many did they deliver altogether?

_____

_____

2 The ratio of students who wish they have X-ray vision to those who don't is 4:5. If 52 students wished they had X-ray vision, how many students were asked?

_____

_____

3 If 38 loaves of white bread were donated to the foodbank, and the ratio of white loaves to wholemeal was 2:7, how many wholemeal loaves were there?

_____

_____

4 Zifa is making a fruit drink using orange, apple and mango juice in the ratio of 3:4:2. If she has only 90 mL of mango juice, what is the maximum amount of fruit drink she can make?

_____

_____

5 **a** Students in Year 9 at Paradise High study Te reo Māori, French and German in the ratio of 5:6:3. If 48 students study German, how many students are in Year 9?

_____

_____

b Discuss any assumptions and limitations of your answer.

_____

_____

6 **a** The ratio of the diameter of a large pizza to that of an extra-large pizza is 8:9. If an extra-large pizza has a diameter of 36 cm, what is the diameter of a large pizza?

_____

_____

b Calculate the ratio of the area of the smaller pizza to that of the larger pizza. If the smaller pizza cost $15.99, how much should the larger pizza cost? State any assumptions of your answer.

_____

_____

_____

_____

 ISBN: 9780170477550

Non-proportional sharing

- Rates and ratios are not always proportional.
- Some cultures share based on need rather than proportional contribution.

Example: Tui is moving into a flat with her friends who are a couple. One of the couple works from home.

> The monthly rent is $1800.
> Power is $155 a month.
> Internet is $95 a month.

Demonstrate some ways in which the bills could be split and justify the reasoning.

Option 1: Weekly rent is split by bedroom and the rest evenly shared.
Tui pays $900 + $83.33 = $983.33
Couple pay $900 + $166.66 = $1066.66
Assuming the couple share a bedroom, they shouldn't have to pay as much as Tui who has her own.

Option 2: Share everything equally.
$2050 ÷ 3 = $683.33 each per month.

While we assume that the couple share a room, if one of them works from home, it is likely that they will use more power and internet.

Answer the following questions.

1 Three neighbours need to replace their fence at a cost of $220 per metre.
Suggest at least two ways they could share the cost.
Justify reasons for your answers.

Neighbour A   Neighbour B

4.2 m   2.1 m

Neighbour C

_____

_____

_____

_____

_____

2 Beau and Buddy got paid $500 to do some painting. Beau is a faster painter and completes 3 square metres to Buddy's 2 square metres. Beau worked for three days and Buddy worked for five. Describe at least two different ways in which they could be paid.

_____

_____

_____

_____

Exchange rates

- The value of the New Zealand dollar changes in relation to other currencies and with time.
- Changes in its value are important when calculating the cost of overseas goods or travel.
- Currency units: the New Zealand dollar can be written as $NZ or NZD.

Use these rates for the calculations in this exercise:

$NZ1 = €0.55 (euro)
$NZ1 = £0.47 (pounds sterling)
$NZ1 = $US0.59 (US dollar)
$NZ1 = $AU0.92 (Australian dollar)

Converting currency

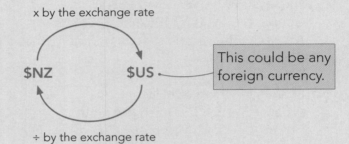

x by the exchange rate

$NZ $US

This could be any foreign currency.

÷ by the exchange rate

Examples:

1 Kaea takes $1500 spending money with him to Australia. How many $AU can he buy?

$NZ1 = $AU0.92
$NZ1500 = $AU1500 x 0.92
= $AU1380

2 Isla came back from a trip to Europe with €460. How many $NZ will she get when she exchanges it?

€0.55 = $NZ1
€460 = $NZ460 ÷ 0.55
= $NZ836.36 (2 dp)

Solve the following exchange rate problems. Round to 2 dp.

1 Convert $NZ3500 to $US.

_____

_____

_____

2 Convert $NZ400 to €.

_____

_____

_____

3 Convert $NZ550 to £.

_____

_____

_____

4 Convert €240 to $NZ.

_____

_____

_____

5   Convert £50 to $NZ.

_____

_____

_____

6   Convert $AU350 to $NZ.

_____

_____

_____

7   Convert $NZ890 to $AU.

_____

_____

_____

8   Convert $US1585 to $NZ.

_____

_____

_____

9   Jemimah received $US200 from her uncle for her birthday. How much is that in $NZ?

_____

_____

_____

10   Rodney returned to New Zealand from a holiday and found he had €65 and £105 left over. How much can he expect to receive in $NZ?

_____

_____

_____

11   Hemi went on holiday and took $AU1150 with him. How many $NZ did he exchange?

_____

_____

_____

12   An average coffee price in New Zealand is $4.35. How does that compare with England where a cup will cost you £3.40?

_____

_____

_____

13   Mia returned from holiday and had leftover cash from the countries she visited. She had €44.00, £314.70 and $AU22.80. How much is this worth in total in New Zealand dollars?

_____

_____

_____

14   Sarah went to Europe and changed $NZ250 into euros when the rate was $NZ1 = €0.55. She spent two thirds of it and exchanged the rest on her return to New Zealand when the rate was €1:$NZ1.75. How much does she bring back in New Zealand dollars?

_____

_____

_____

Simple interest

- When you deposit money into the bank, the bank pays you '**interest**' for the use of that money.
- When you borrow money, you have to pay '**interest**' because you are using the bank's money.
- Simple interest is the same amount paid to you or by you at regular intervals.

Examples:

1 Piri deposits $9500 into a savings account. The bank will pay 3% interest annually.
How much interest does he earn after a year?

$$3\% \text{ of } \$9500 = 0.03 \times 9500$$
$$= \$285$$

2 Maise borrows $2100 off her parents to buy a computer. Maise will pay them 5% of $2100 in interest each year.

  a How much interest does she pay each year?

$$5\% \text{ of } \$2100 = 0.05 \times 2100$$
$$= \$105$$

  b If it takes her three years to pay the $2100 back, how much interest will she have paid her parents in total?

$$3 \times \$105 = \$315$$

Answer the following questions. All interest rates are annual.

1 Complete the table.

Value of loan	Rate of interest	Interest per year	Number of years	Total debt if none paid back
$3000	4%	$120	5	$3000 + ($120 × 5) = $3600
$400	3%		4	
$17 000	5.5%		3	
$259 000	9.7%		8	

2 Calculate the simple interest earned each year when $9000 is invested at 4%.

_____

3 Calculate the simple interest earned when $3700 is invested at 7% for three years.

_____

4 **a** Lucy borrows $5000 from her brother to buy a motorbike. Lucy paid him 5% in interest each year. How much does she pay each year?

_____

  b If she pays all the money back after four years, how much will the motorbike have cost her in total?

_____

5 **a** Ula invested her $12 000 in a savings account at a 3.5% interest rate. How much interest will she earn each year? _____

  b How long will she need to leave her money in there to earn $1500 in interest?

_____

 ISBN: 9780170477550

Compound interest

- With compound interest, the interest is added to the amount deposited.
- In the following period, interest is calculated on the **total** (deposit and previous year's interest).

Formula if interest applied annually

$$A = P(1 + r)^t$$

Formula if interest is applied more often than annually

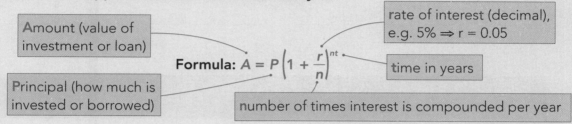

Amount (value of investment or loan)

rate of interest (decimal), e.g. 5% $\Rightarrow$ r = 0.05

Formula: $A = P\left(1 + \dfrac{r}{n}\right)^{nt}$

time in years

Principal (how much is invested or borrowed)

number of times interest is compounded per year

Examples:

1 Luka borrows $3500 from her parents to buy a car. Her parents charge her 4% per year compound interest. This means that she doesn't pay anything back annually, but pays back the $3500 plus compound interest after three years. How much would she pay back at the end of three years?

End of the first year she owes $3500 + $3500 x 4% = $3640.00
End of the second year she owes $3640 + $3640 x 4% = $3785.60
End of the third year she owes $3785.60 + $3785.60 x 4% = $3937.02 (2 dp)

Or $3500 x 1.04 x 1.04 x 1.04 = $3937.02 (2 dp)

Or $3500 x 1.04^3 = $3937.02 (2 dp)

2 Noa invested $6000 into a savings account with annual compound interest for five years. At the end of 5th year, he had $7299.92 in his account. What was the compound interest rate for this account?

$$A = P(1 + r)^t$$
$$\$7299.92 = \$6000 \times (1 + r)^5 \qquad (\div 6000 \text{ and } \sqrt[5]{\ })$$

$$\sqrt[5]{\frac{7299.92}{6000}} = (1 + r)$$

$$1.04 = 1 + r$$

$$r = 0.04 \Rightarrow \text{the rate was 4\%}$$

3 Lachlan invests $5800 in an account that **compounded monthly** with a rate of 3%. How much would his investment be worth at the end of four years? Describe any assumptions in your answer.

$$A = P\left(1 + \frac{r}{n}\right)^{nt}$$

$$= \$5800\left(1 + \frac{0.03}{12}\right)^{12 \times 4}$$

$$= \$6538.50$$

Assumptions: We assume Lachlan didn't withdraw any money over the four years and that the rate of interest remained constant over this time period.

Answer the following questions.

1 Complete the table. All these rates compound annually.

Value of loan	Rate of interest	Number of years	Calculation	Total debt
$3000	4%	5	3000×1.04^5	$3649.96
$400	3%	4		
$17 000	5.5%	3		
$259 000	9.7%	8		

2 Complete the table.

Value of loan	Rate of interest	Number of years	Compound period	Calculation	Total debt
$3000	4%	5	Every six months	$3000 \left(1 + \dfrac{0.04}{2}\right)^{2 \times 5}$	$3656.98
$400	3%	4	Monthly		
$17 000	5.5%	3	Fortnightly		
$259 000	9.7%	8	Weekly		

3 $4500 is invested at 5% compound interest, paid annually. Calculate its value at the end of two years.

4 $12 000 is invested at 7.5% compound interest, paid annually. Calculate its value at the end of three years.

5 $24 000 is invested with compound interest paid annually. At the end of three years, it was worth $26 609.23. What was the compound interest rate?

6 At the end of four years, Hana had $7524.56. She had invested her money in an account with a compound interest of 5% paid annually for three years. How much was her original investment?

7   Tui has $11 000 to invest. She has two offers: scheme one is 4.2% simple interest; scheme two is 4% compound interest paid annually.

a   Complete the table to show how much Tui has after one, two, three and four years.

Value of investment after ...	Scheme one	Scheme two
one year	$11 462.00	$11 440.00
two years		
three years		
four years		

b   Which scheme do you suggest Tui invests in? Justify your answer clearly, and explain any potential effects or limitations of her choices.

_____

_____

_____

8   Susie has $25 000 to invest. She has two opportunities: scheme one is 3.5% compound interest paid annually; scheme two is 3.5% compound interest, paid every six months.

a   If she withdraws her investment at the end of four years, how much will each scheme yield?

Scheme one: _____

_____

_____

Scheme two: _____

_____

_____

b   What interest rate would Susie require on scheme one to give a better result than scheme two at the end of three years?

_____

_____

_____

_____

c   Describe any assumptions you have made and how they might affect your answer.

_____

_____

_____

Measurement fundamentals

Unit conversion

Use the following charts to help you convert units.

Length

Mass

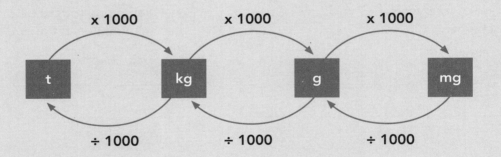

Capacity (fluids)

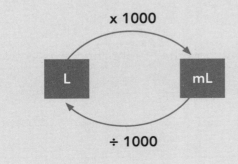

 ISBN: 9780170477550

Area

- Length is measured in mm, cm, m or km.
- Area is measured in mm², cm², m², ha (hectares) or km².

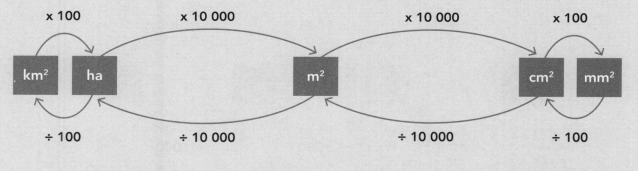

$$1\ km^2 = 100\ ha \qquad 1\ ha = 10\ 000\ m^2 \qquad 1\ m^2 = 10\ 000\ cm^2 \qquad 1\ cm^2 = 100\ mm^2$$

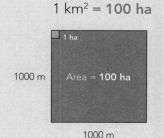
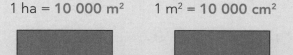
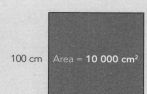
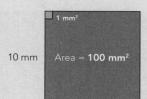

Example: Convert 984 100 cm² into m².

$$1\ cm^2 = 0.0001\ m^2$$

∴ 984 100 cm² = 984 100 cm² ÷ 10 000

= 98.41 m²

> Your answer will be a smaller number, so **divide**.

Convert the following.

1. 700 m² = _____

    = _____ ha

2. 3500 ha = _____

    = _____ km²

3. 96 500 cm² = _____

    = _____ m²

4. 7.1 m² = _____

    = _____ cm²

5. 1070 mm² = _____

    = _____ cm²

6. 25 980 m² = _____

    = _____ km²

7. 0.031 ha = _____

    = _____ m²

8. 129 750 mm² = _____

    = _____ m²

9. A section of land is 123 m long and 135 m wide. Calculate its area in ha.

    _____

    _____

10. The floor plan of a rectangular house 183m². If it is 1906 cm long, how wide is it?

    _____

    _____

Volume

- Volume is measured in mm^3, cm^3, mL, L or m^3.
- cm^3 are often used in medical contexts, but they are called '**ccs**' (**c**ubic **c**entimetres).

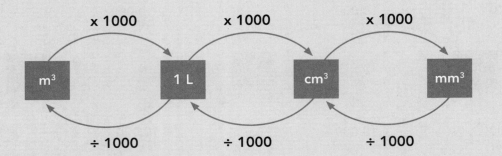

x 1000   **x 1000**   **x 1000**

m^3 1 L cm^3 mm^3

÷ 1000   **÷ 1000**   **÷ 1000**

$1 cm^3 = 1 mL = $**$1000 mm^3$** $1 L = 1000 mL = $**$1000 cm^3$** $1m^3 = 1\ 000\ 000\ cm^3$
$= 1000\ L$

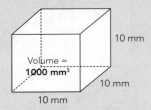

Volume =
$1000\ mm^3$
10 mm
10 mm
10 mm

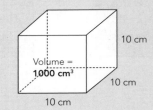

Volume =
$1000\ cm^3$
10 cm
10 cm
10 cm

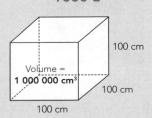

Volume =
$1\ 000\ 000\ cm^3$
100 cm
100 cm
100 cm

Example: Convert $0.84\ cm^3$ into mm^3.

$$1\ cm^3 = 1000\ mm^3$$
$$\therefore\ 0.84\ cm^3 = 1000 \times 0.84\ mm^3$$
$$= 840\ mm^3$$

Convert the following.

1 $460\ 000\ cm^3 = $ _____

  $= $ _____ m^3

2 $87\ 600\ mm^3 = $ _____

  $= $ _____ cm^3

3 $0.0071\ m^3 = $ _____

  $= $ _____ cm^3

4 $0.0451\ cm^3 = $ _____

  $= $ _____ mm^3

5 $1.5\ L = $ _____

  $= $ _____ cm^3

6 $26\ 700\ cm^3 = $ _____

  $= $ _____ L

7 $0.0082\ m^3 = $ _____

  $= $ _____ L

8 $160\ 000\ cm^2 = $ _____

  $= $ _____ m^3

9 A cuboidal tank is 133 cm by 70 cm and is 0.42 m deep. How many litres of water does it contain when full?

_____

10 How many $200\ cm^3$ glasses can be filled from a 2.5 L bottle of juice?

_____

Speed

- Speed is usually measured in metres per second (m/s) or kilometres per hour (km/h).

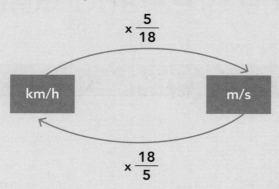

Example:	Usain Bolt has run 100 m in 9.58 seconds. Using this information, how many km per hour could he run?	
Step 1:	Calculate how many metres per second. At this pace, he is running 10.44 m per second.	$100 \div 9.58 = 10.44$ (2 dp)
Step 2:	Convert 10.44 metres into km. He is running 0.01044 km per second.	$10.44 \div 1000 = 0.01044$ (5 dp)
Step 3:	Convert seconds to hours (3600 seconds in an hour). Usain could run 37.58 km per hour.	$0.01044 \times 3600 = 37.58$
Limitation:	He can run at this speed for 100m, but it is unlikely that he could maintain it for longer distances.	

State whether these are true or false. If false, then calculate the correct conversion.

1 **a** 72 km/h = 20 m/s

      b 30 m/s = 108 km/h

      c 6 km/min = 1000 cm/s

      d 52 m/min = 31.2 cm/s

      e 85 km/h = 2361.11 cm/s

      f 105 cm/min = 0.63 km/h

2 Some of the fastest freestyle swimmers can cover 50 m in 22 seconds. At this rate, how long would it take them to swim 1 km? State any limitations of your answer.

Three-dimensional shapes: surface area and volume

Complete the table to make your own list of formulae.

Shape	Volume	Surface area
Triangular prism		
Cylinder		
Cone		
Pyramid		
Sphere		

 # Geometry fundamentals

Theorem of Pythagoras

- The Theorem of Pythagoras applies to **right-angled** triangles only.
- The longest side is the **hypotenuse**.
- The hypotenuse is always **opposite** the right angle.
- The theorem is used for finding **lengths** of sides.

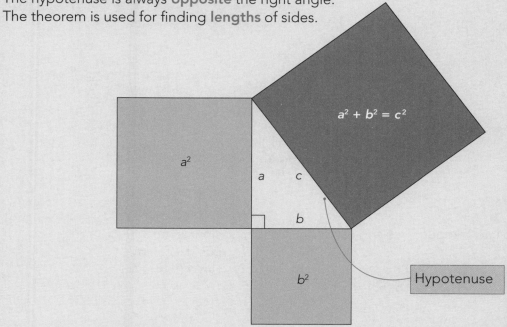

$$\text{short side}^2 + \text{short side}^2 = \text{hypotenuse}^2$$
$$a^2 \quad + \quad b^2 \quad = \quad c^2$$

Examples:

1 Finding the length of the hypotenuse
Calculate the length of c.

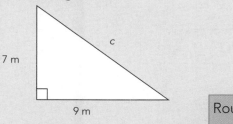

$$7^2 + 9^2 = c^2$$
$$\sqrt{7^2 + 9^2} = c$$
$$c = \sqrt{130}$$
$$c = 11.4 \text{ m (1 dp)}$$

Round your answers to one more decimal place than the lengths you are given.

11.4 m is a reasonable answer because it's longer than the other two sides.

2 Finding the length of short sides
Calculate the length of x.

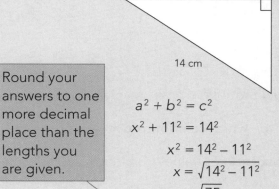

$$a^2 + b^2 = c^2$$
$$x^2 + 11^2 = 14^2$$
$$x^2 = 14^2 - 11^2$$
$$x = \sqrt{14^2 - 11^2}$$
$$x = \sqrt{75}$$
$$x = 8.7 \text{ cm (1 dp)}$$

8.7 cm is a reasonable answer because it's shorter than the hypotenuse.

21

Trigonometry

- Trigonometry can also be used for calculations involving **right-angled triangles**.
- However, unlike calculations using the Theorem of Pythagoras, an **angle must** be involved.

Before you begin a calculation involving trigonometry, **label** the sides of the triangle:

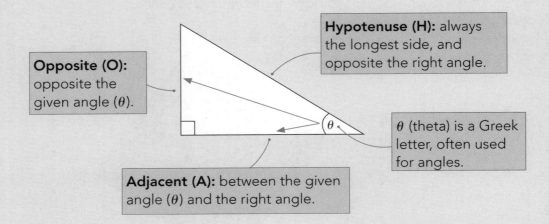

Opposite (O): opposite the given angle (θ).

Hypotenuse (H): always the longest side, and opposite the right angle.

θ (theta) is a Greek letter, often used for angles.

Adjacent (A): between the given angle (θ) and the right angle.

SOH CAH TOA

You will need to know about three trigonometric functions: **sin** θ (sine)
cos θ (cosine)
tan θ (tangent).

These are the rules in trigonometry:

$$\sin \theta = \frac{O}{H} \qquad \cos \theta = \frac{A}{H} \qquad \tan \theta = \frac{O}{A}$$

These are usually remembered as the 'word' **SOH CAH TOA**.

Organising **SOH CAH TOA** into triangles can help you to work out how to use it:

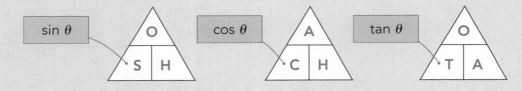

Process

Step 1: **Label** the sides that are involved in the problem with O, A and H.

Step 2: Decide **which relationship** involves these sides.

Step 3: **Substitute** the values from the triangle.

Step 4: **Think about your answer — does it seem sensible?**

Important: Check your calculator is in degrees.

ISBN: 9780170477550

Examples:

1 Finding a short side

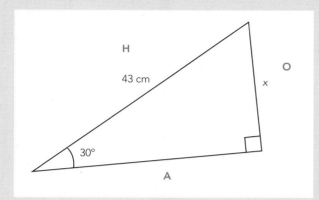

$$\sin \theta = \frac{O}{H}$$

$$\sin 30° = \frac{x}{43}$$

$$x = 43 \times \sin 30°$$

$$= 21.5 \text{ cm (2 dp)}$$

> 21.5 cm seems a reasonable answer because it's shorter than the hypotenuse.

2 Finding a long side

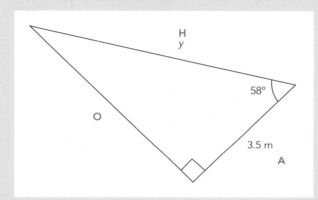

$$\cos \theta = \frac{A}{H}$$

$$\cos 58° = \frac{3.5}{y}$$

$$y \times \cos 58° = 3.5$$

$$y = \frac{3.5}{\cos 58°}$$

$$y = 6.60 \text{ m (2 dp)}$$

> 6.60 m seems a reasonable answer because it's the hypotenuse and it's longer than the adjacent side.

3 Finding an angle

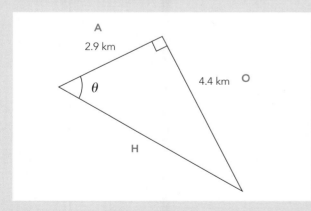

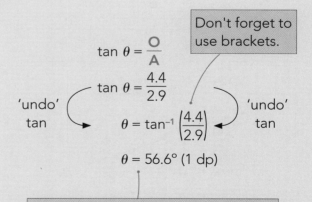

> Don't forget to use brackets.

$$\tan \theta = \frac{O}{A}$$

$$\tan \theta = \frac{4.4}{2.9}$$

'undo' tan $\theta = \tan^{-1}\left(\frac{4.4}{2.9}\right)$ 'undo' tan

$$\theta = 56.6° \text{ (1 dp)}$$

> Think about your answer. θ is opposite the longer of the two perpendicular sides, so an answer between 45° and 90° is expected.

Use trigonometry to calculate the unknown lengths and angles in each triangle. Give your answers for lengths to 4 sf and angles to 1 dp.

1

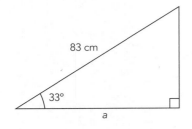

83 cm

33°

a

_____

_____

_____

2

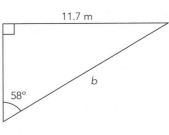

11.7 m

58°

b

_____

_____

_____

3

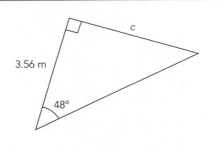

c

3.56 m

48°

_____

_____

_____

4

382 mm

41°

d

_____

_____

_____

5

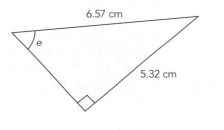

6.57 cm

e

5.32 cm

_____

_____

_____

6

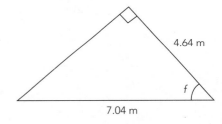

4.64 m

7.04 m

f

_____

_____

_____

7

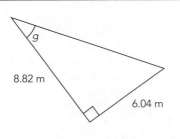

g

8.82 m

6.04 m

_____

_____

_____

8

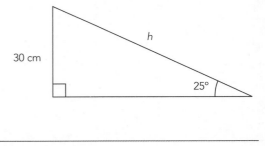

h

30 cm

25°

_____

_____

_____

9

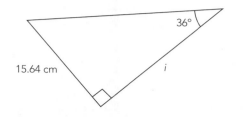

10

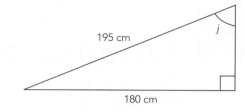

_____

_____

_____

_____

_____

_____

11

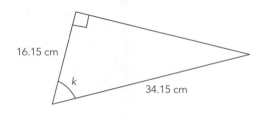

12 Calculate the length of the diagonal of this square.

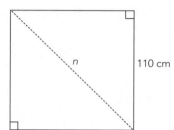

_____

_____

_____

_____

_____

_____

13 Calculate the dimensions _p_ and _q_ of this rectangle.

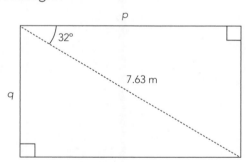

_____

_____

_____

_____

_____

_____

_____

_____

14 Calculate the angle _s_ and the vertical height _t_ of this parallelogram.

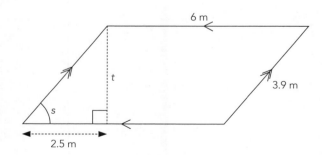

_____

_____

_____

_____

_____

_____

_____

_____

Optimal solutions

- Finding an optimal solution for a situation means finding a feasible solution that reaches a **maximum** or **minimum** value.
- Examples:
  — the most profit
  — the best value for money
  — the least cost.
- Sometimes this involves thinking beyond just the mathematics of the situation.
- There isn't necessarily **one** correct answer. There could be a range of sensible solutions.
- Limitations and assumptions should be considered for each situation.

Example: Pots of yogurt

In order to be called 'yogurt', fruit yogurt must contain at least 80% yogurt.
Yogurt costs $5.88 per kg and fruit costs $10.52 per kg.
Empty 125 g pots cost 12c each and 200 g pots cost 18c each.
125 g pots of yogurt sell for $1.50 each and 200 g pots sell for $2.40 each.
Tama can make batches of 20 kg of yogurt at a time and he has unlimited fruit available.
Which size of pot will maximise his profit from each 20 kg batch of yogurt?
Discuss any assumptions and limitations to your answer.

For yogurt containing the minimum amount of yogurt and maximum fruit
Profit from a 125 g pot:

80% yogurt $\Rightarrow$ ratio of fruit to yogurt = 1:4 $\therefore$ a 125 g pot needs ratio of 25 g:100 g

$$\text{Cost of fruit} = \frac{25}{1000} \times \$10.52 \qquad \text{Cost of yogurt} = \frac{100}{1000} \times \$5.88$$

$$= \$0.26 \text{ (2 dp)} \qquad\qquad = \$0.59 \text{ (2 dp)}$$

Total cost of pot contents = $0.26 + $0.59 + $0.12 = $0.97

$$\text{Profit from one 20 kg batch of yogurt} = \frac{20}{0.125} \times (1.50 - 0.97)$$

$$= \$84.80$$

Profit from a 200 g pot:

Ratio of fruit to yogurt = 1:4 $\therefore$ a 200 g pot needs ratio of 40 g:160 g

$$\text{Cost of fruit} = \frac{40}{1000} \times \$10.52 \qquad \text{Cost of yogurt} = \frac{160}{1000} \times \$5.88$$

$$= \$0.42 \text{ (2 dp)} \qquad\qquad = \$0.94 \text{ (2 dp)}$$

Total cost of pot contents = $0.42 + $0.94 + $0.18 = $1.54
$$\text{Profit from one 20 kg batch of yogurt} = \frac{20}{0.2} \times (2.40 - 1.54)$$

$$= \$86.00$$

Conclusion: There is $1.20 more profit to be made using 200 g pots, which very small. Therefore which size Tama chooses is likely to depend on other factors such as costs of labour, packaging, customer preference, etc.

Assumptions and limitations: Assumed that the costs of packing and transport are the same for both sized pots. It may be that customers would prefer a greater ratio of yogurt to fruit, or have a preference for the size of pot, in which case the profit would change.

 ISBN: 9780170477550

Swimming pool memberships

Fergus is keen to start swimming this year. He has a number of membership options to consider.

Option one	Option two	Option three
12-month membership $719.80 Unlimited entry	$6.70 per visit	20-visit card $120.60 Bring a friend for $4.50

1　**a**　If Fergus decided to buy a 12-month membership, how many times would he need to visit the pool for it to be cheaper than option two?

　　　　_____

　　　　_____

　　　　_____

b　How many times each week would he need to swim? Show your calculations.

_____

_____

_____

_____

2　If Fergus decided on option three and took a friend each visit, how much would they save compared to option two?

_____

_____

_____

_____

3　If Fergus wants to swim three times a week, which membership option would you recommend? Justify your answer clearly, and explain any potential effects or limitations of his choices.

_____

_____

_____

_____

_____

_____

_____

Wildlife park

A wildlife park charges the following prices:

Daily charges		Annual pass (unlimited entry valid for 12 months)	
Adult	$34.50	1 adult and 1 child (fixed child, any adult)	$118.00
Senior or student	$28.00	Adult annual (fixed adult)	$82.00
Child	$13.00	Child annual (fixed child)	$28.00
Under 5 years	Free		
Family pass (2 adults and 2 children)	$85.00		
Conditions:	Fixed means that the person must be named and ID produced on request. Child must be aged between 5 and 15 years old. If an annual pass is renewed, then there is a 10% discount.		

1 **a** If a family visit the park and pay for two adults and three children, how much will it cost them?

_____

_____

b How much would they save by buying a family pass?

_____

_____

2 Eight-year-old Janine and her grandfather (a senior) want to visit the wildlife park regularly. Which option do you recommend for them? Justify your answer clearly, and explain any potential effects or limitations of their choices.

_____

_____

_____

_____

_____

_____

_____

_____

_____

_____

_____

_____

_____

ISBN: 9780170477550

Sunscreen

Ronnie is going on a five-day holiday and wants to work out how much sunscreen to take. Sunscreen should be applied every three hours during the day.

Every square centimetre of exposed skin should have approximately 0.002 millilitres of sunscreen applied to it.

The Du Bois formula calculates the surface area (m^2) of the human body, where W is mass (kg) and H is height (cm):

$$SA = 0.007184 \times W^{0.425} \times H^{0.725}$$

Ronnie weighs 78 kg and is 175 cm tall.

Investigate how much sunscreen you recommend Ronnie takes on holiday. Discuss any assumptions and limitations.

Wheelie bins

Justin is trying to decide if he should upgrade the size of his organics wheelie bin. The council empties it each week. The cost for an 80 L bin is included in his rates, but for an additional annual fee of $220 he can upgrade to a 240 L bin.

The local dump charges by weight at a rate of $131.25 per tonne for organic waste, with a minimum charge of $14.90.

Compost/green waste weighs approximately 0.033 kg per 1 L.

Compare and comment on Justin's options. Discuss any assumptions or limitations of your answer.

240 L   80 L

Spa pool

Chloe has just purchased a spa pool that is based on a regular octagon which is extended to create a right angle as shown. The diagram shows the internal dimensions. She needs to chlorinate the water.

It is recommended that 1.5 teaspoons of granules per 1000 litres of water per week should be added to a spa to keep it free from bacteria.

1 teaspoon = 5 mL = 5 grams

The maximum depth of the spa is 1.37 m.

Investigate the amount of chlorine Chloe should purchase for her spa for a year. It comes in 1 kg, 2 kg and 5 kg bags.

Discuss any assumptions and limitations.

80 cm

_____

_____

_____

_____

_____

_____

_____

_____

_____

_____

_____

_____

_____

_____

_____

_____

_____

_____

_____

Choco bars

Due to inflation, William needs to rethink the size and
shape of his popular-selling Choco bar in order to
minimise the cost of ingredients.

- The bar is currently cuboidal (like a box) in shape with
  dimensions of 12 cm x 3.51 cm x 1.78 cm and it weighs
  75 grams.
- He wants to keep the length (12 cm) the same but
  reduce the mass by 16%.
- His machinery can make only cuboidal chocolate bars.
- The chocolate coating is 2 mm thick and the chocolate
  is more expensive than the filling.

Suggest some ways in which William could change the chocolate bar design in order to minimise
the cost of the chocolate needed. Justify your suggestions using calculations to 2 dp. Discuss any
assumptions and limitations you have made.

Hint: It might be helpful to create a table.

    ISBN: 9780170477550

He is investigating upgrading his equipment so that he could make the Choco bar with a cylindrical shape. Assuming he keeps the length at 12 cm, investigate whether this would enable him to reduce the chocolate needed for each bar.

_____

_____

_____

_____

_____

_____

_____

_____

_____

_____

_____

_____

_____

_____

_____

_____

_____

_____

_____

_____

_____

_____

_____

_____

_____

Practice tasks

Practice task one

You are to compose a report on e-scooters. Your report will be divided into two parts.
Section A is based on the e-scooter ACC claims, and Section B is based on renting versus buying an e-scooter.
You will use the following questions to form the basis of your report.
You must refer to specific information contained in the graphs throughout your report.

Section A: E-scooter ACC claims

- Compare and comment on the ages of people who are making ACC claims for e-scooter injuries.
- Comment on where and when these injuries are most likely to occur, and the natures of these injuries. Discuss any assumptions of your answer.
- News articles are suggesting that the number of e-scooter injuries is increasing. Comment on whether these claims can be supported. Give at least TWO different supporting comments. Justify and support your conclusions using resources C and D.
- Using the information in the resources and your knowledge of e-scooters, how could the number of e-scooter injuries be reduced in New Zealand? Discuss any assumptions and limitations of your answer.

Section A resources

All information on this page has been sourced from the Accident Compensation Corporation (ACC) and it all relates to e-scooter injuries.

Note: 2023 data is YTD (Jan–Oct).

A

Age group	Total claims
0–9	44
10–19	567
20–29	624
30–39	460
40–49	270
50–59	224
60+	72

B

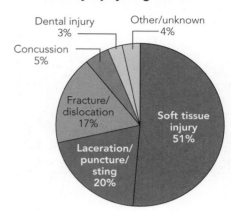

Primary injury diagnosis 2023

C

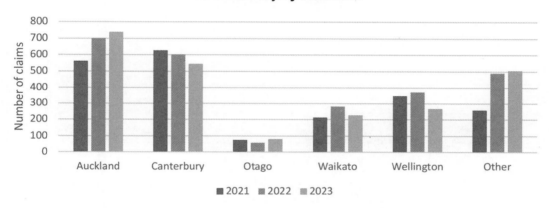

E-scooter injury locations

D

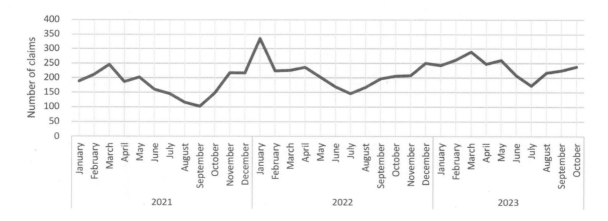

Number of new claims for e-scooter injuries

Section B: Renting versus buying an e-scooter

- Riley has a 12-minute commute to and from work five days a week. Compare and comment on the brand of e-scooter that would be best for him to use. Justify your answer using the information in resources E, F and G.
- With the information provided in resource H, compare options if Riley is considering buying a scooter that he can only charge at home. Discuss any assumptions and limitations of your answer.

_____

_____

_____

_____

_____

_____

_____

_____

_____

_____

_____

_____

_____

_____

_____

_____

_____

_____

_____

_____

_____

_____

_____

_____

_____

  ISBN: 9780170477550

Section B resources

E

E-scooter rental options

There are three rental companies — their prices are in the table below.

	Lemon	Beat	Synapse
Unlock fee	$0.45	$1.00	$AU1.00
Cost per minute	$0.45	$0.40	$AU0.45
Weekly pass	$55.99 (unlimited time)	$49.99 (max 20 min/day)	$AU50.00 (one hour/day)
Extra	Helmet provided		Helmet provided

F

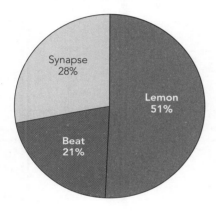

Number of rental e-scooters

Synapse 28%
Lemon 51%
Beat 21%

G

Exchange rates

$NZ1 = $US0.59 (US dollar)
$NZ1 = $AU0.92 (Australian dollar)

H

E-scooter prices

	Cloudy Rides	Beat	E-Motion
Price	$US650	$599.99	$1399.00
Top speed	30 km/h	25 km/h	40 km/h
Range	35 km	40 km	45 km
Other	Foldable High water resistance	Non-foldable Second-hand	Non-foldable

We pay the GST!

ISBN: 9780170477550

Practice task two

You are to compose a report on the cost of living and food waste in New Zealand. Your report will be divided into two parts.

Section A is based on the cost of living, and Section B is based on food waste.

You will use the following questions to form the basis of your report.

You must refer to specific information contained in the graphs throughout your report.

Section A: Cost of living

The cost of living was a major concern of New Zealanders in 2023.

- From the information in resource A, rank the four biggest household expenses for those accessing community support from largest to smallest. Justify your decisions with reference to the information provided in resource A.
- Comment on the methodology of the data in resource C. Consider sample method, size and who the findings could be confidently applied to.
- In September 2023, a Stuff headline read 'Meltdowns, dumped trolleys: rise of food price anxiety in New Zealand'. Give at least TWO different supporting comments. Justify and support your conclusions using information from resources B and C.

_____

_____

_____

_____

_____

_____

_____

_____

_____

_____

_____

_____

_____

_____

_____

_____

_____

_____

_____

_____

_____

_____

_____

_____

Section A resources

A

Results from a survey of 149 people who recently access community food support, showing the first, second and third most costly expenses.

For instance: for 69.80% of household, housing was their greatest expense, 34.23% of households found that kai was their second highest expense and 35.57% of households found kai was their third highest expense.

Highest household expenses

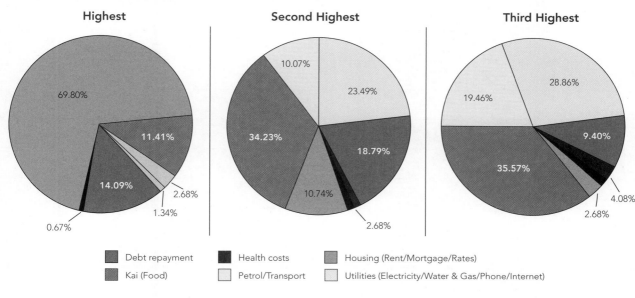

| Debt repayment | Health costs | Housing (Rent/Mortgage/Rates) |
| Kai (Food) | Petrol/Transport | Utilities (Electricity/Water & Gas/Phone/Internet) |

B

These are the results from a survey of 1000 New Zealanders done during 2023 by Consumer NZ.

Consumers' biggest financial concern in the next year is the cost of groceries

Q. Which of the following are your biggest financial concerns over the next 12 months? The graph shows the proportion of people who ranked each concern in the top three. Rent and mortgage repayments not displayed.

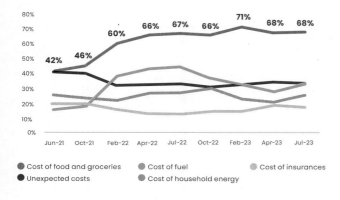

- Cost of food and groceries
- Unexpected costs
- Cost of fuel
- Cost of household energy
- Cost of insurances

C

Overview of methodology

Kantar conducted 1501 online interviews between 25 July and 14 August 2023. Data collected was nationally representative to ensure that results could be used to measure New Zealanders' attitudes and behaviours.

Household food spend (average week, 2023)

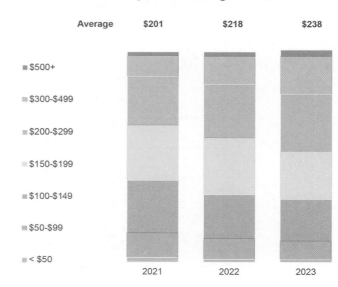

Section B: Food waste

- Using resource D, write a description of the group of people who waste the most food. Provide numerical evidence to support your description.
- It is claimed that there has been almost no change in the numbers of people disposing of food scraps in compost bins or worm farms. Comment on this claim using the information provided in resource E.
- 'Better food education in schools could slash food waste' — is this headline justified? If so, using information supplied in resource F, suggest some topics that should be included in the school curriculum.

_____

_____

_____

_____

_____

_____

_____

_____

_____

_____

_____

_____

_____

_____

_____

_____

_____

_____

_____

_____

_____

_____

_____

_____

_____

 ISBN: 9780170477550

Section B resources

The following resources come from the Rabobank KiwiHarvest New Zealand Food Waste Survey 2023 results.

xx/xx *Significantly **higher**/**lower** than total*   ▲▼ *Significantly **higher**/**lower** than previous year*

D

New Zealanders estimate that 12.2% of their annual household food spend goes to waste.

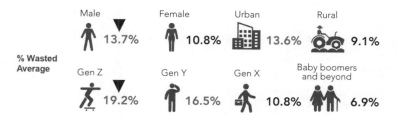

% Wasted Average

Male 13.7% ▼	Female 10.8%	Urban 13.6%	Rural 9.1%
Gen Z 19.2% ▼	Gen Y 16.5%	Gen X 10.8%	Baby boomers and beyond 6.9%

This translates to **$1,510 per household** per year of food wasted or **$3.2 billion of waste each year** [1]

$3.1b in 2022
$2.4b in 2021

E

Actions taken to reduce food waste (2023)

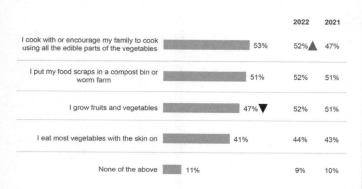

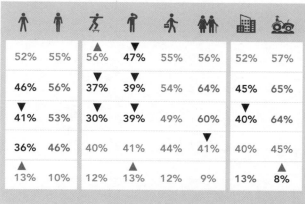

	2023	2022	2021
I cook with or encourage my family to cook using all the edible parts of the vegetables	53%	52% ▲	47%
I put my food scraps in a compost bin or worm farm	51%	52%	51%
I grow fruits and vegetables	47% ▼	52%	51%
I eat most vegetables with the skin on	41%	44%	43%
None of the above	11%	9%	10%

🚹		🛹	🧍	🚶	👥	🏢	🚜
52%	55%	56% ▲	47% ▼	55%	56%	52%	57%
46%	56%	37% ▼	39% ▼	54%	64%	45%	65%
41% ▼	53%	30% ▼	39% ▼	49%	60%	40% ▼	64%
36%	46%	40%	41%	44%	41% ▼	40%	45%
13% ▲	10%	12%	13% ▲	12%	9%	13%	8% ▲

F

Key reasons for wasted food

> Appropriate food storage knowledge, and not knowing what to do with leftovers are common reasons for food waste for Gen Z and Y
>
> On top of that, there are more Gen Z who are not preparing food properly and not eating it as a result and not able to store food properly.
>
> Kids not finishing their plate is significant among Gen Y and X

	Food going off before you can finish it	Food going off before its 'use buy' or 'best before' date	Not planning sufficiently	Food not being as good as you expected it to be when you bought it	Children not eating food that is prepared for them	Not being able to finish what goes on to your plate	Buying too much	Not knowing what to do with leftovers	Other (please specify)	Not preparing food properly and not eating it as a result	We aren't sure how to store food properly
2023	50%	32%	21%	20%	18%	18% ▲	17%	8%	7%	7% ▲	3%
2022	50% ▼	34% ▲	20% ▲	23% ▲	16%	15%	18%	8%	8% ▲	5%	3% ▲
2021	63%	28%	16%	18%	16%	17%	16%	8%	6%	6%	1%

Section C: Food plans

- Using resource G, write a description of how the use of food delivery businesses has changed over the last three years.
- A family of four is investigating signing up to a food plan for a month. Each week, these deliver to their home a box or bag containing all the ingredients and recipes needed for three, four or five evening meals. Using resource H, make a recommendation to the family of four. Justify your answer clearly, and explain any limitations to the choices and any assumptions that have been made.

Section C resources

G

Usage of food delivery services (used last 12 months, 2023)

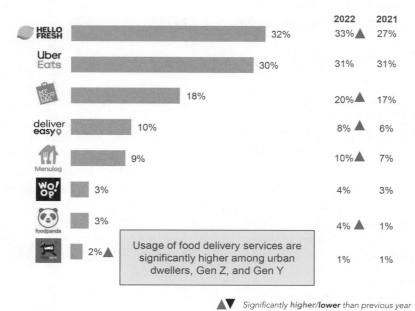

		2022	2021
HELLO FRESH	32%	33% ▲	27%
Uber Eats	30%	31%	31%
MY FOOD BAG	18%	20% ▲	17%
deliver easy	10%	8% ▲	6%
Menulog	9%	10% ▲	7%
WOOP	3%	4%	3%
foodpanda	3%	4% ▲	1%
	2% ▲	1%	1%

Usage of food delivery services are significantly higher among urban dwellers, Gen Z, and Gen Y

▲▼ *Significantly **higher**/**lower** than previous year*

H

My Food Box

Plan 1	Plan 2	Plan 3
Three meals for four people $128.00	Four meals for four people $158.00	Five meals for four people $175.00
All plans $9.99 per week for shipping.		
This month's deal: 50% off for the first week, 30% off the second and 10% off the third (not including shipping).		

Dinner Done

Plan 1	Plan 2	Plan 3
Three meals for four people $130.80	Four meals for four people $167.20	Five meals for four people $189.20
Shipping included.		
This month's deal: Take $2 off the price of each serving!		

Answers

Throughout this book your answers may be different from these. If so, check with your teacher.

Statistical investigations (pp. 5–21)
Census and sample (p. 5)

1	Sample	• Accuracy of this information is not important. • The outcome is not very important. • The cost of a census would not be justified.
2	Census	• The outcome matters to New Zealanders so everybody should have a say. • Accuracy is important so that a democratic conclusion can be reached.
3	Sample	• Favourite crisp flavour is important only to the company that produces them. • The cost of a census would not be justified.

Sampling and bias (p. 6)

1 a Biased sample: Black dogs 2
                                White dogs 7
      Unbiased sample: Black dogs 4
                                White dogs 5

b The second sample is biased because the numbers in the sample ~~are/~~are not proportional to the population.

2 a Only the ripe apples are likely to fall off. The unripe apples form part of the population, and they will not be fairly represented in the sample.

b The first 100 students through the gate are likely to be more compliant than those that arrive later, and particularly those who arrive late. Also absentees and students who are justified in arriving late will be under-represented in the sample.

c Many within New Zealand do not own cellphones, particularly the elderly who are more likely to use a landline.
The nature of some people's jobs mean they will not be able or willing to answer their phone between 8 am and 6 pm.

d Kauri trees within 10 m of a track, particularly those really close to it, are likely to have their roots affected by the creation of and continued walking on the track.
The track may also affect the supply of water to the roots.

Sampling methods (pp. 7–9)

1 a Convenience and Self-selected
Justification: Convenience because the supermarket is a smaller location than the entire mall. Self-selected because people can easily avoid filling in the questionnaire.

b Non-representative
Reason: Anyone who does not go to the supermarket at the time of the questionnaire does not have any chance of being selected, and people in the supermarket who don't want to fill it in can easily avoid it. So the group selected is not representative of all the shoppers in the mall.

2 a Simple random
Justification: Every stick has an equal chance of being selected.

b Representative
Reason: Provided all employees wrote their names on a stick and it was put into the container, every employee would have an equal chance of being selected.

3 a Stratified
Justification: The two teams form different groups (strata) in the population and then the number of ākonga randomly chosen from each sport was proportional to the numbers in each stratum.

b Representative
Reason: Within each group the players were randomly selected.

4 a Cluster and Convenience
Justification: Cluster because the vertical form classes are groups that are likely to contain a cross-section of ages, thus forming clusters within the population. Convenience because they were the clusters nearest the administration block.

b Representative
Reason: Vertical form classes should contain a cross-section of ages and abilities.

5 a Systematic
Justification: Every tenth person on the school roll is selected and this is a systematic method.

b Representative
Reason: Every person in the school has an equal chance of being selected.

6 a Convenience and Self-selected
Justification:

b Non-representative
Reason: Anybody who didn't attend the

assembly and absentees will not have the opportunity to be selected. Those who want to stay behind are likely to be students who are trying to avoid going to class, and these will not be typical of all the students in the school.

Collection of data (p. 10)

1 We want to know the person's height alone, not the person's height plus the thickness of their shoes and the height their hair stands up on top of their head.

2 If the height is measured according to a line that slopes down from the top of the head to the scale, the height read off the scale will be too short. And if the line slopes up, then the height read will be too tall.

3 If a set of instructions which includes removal of shoes, flattening of hair and, for instance, using a book held vertically from the top of the head to ensure measurements are read at right angles, then the data should be reliable.

Data collection processes (pp. 11–13)

1 Survey, because kaiako were given a questionnaire with several questions.

2 Poll, because students were asked one question which asked for their opinion.

3 Experimental, because the investigator can control whether sunflowers plants get extra water or not.

4 Survey, because employees were given a questionnaire with several questions.

5 Observational, because the investigator has no control on the results of sports games and can only record the results.

6 Observational, because the investigator can only observe the prices of fruit and has no control of the price.

7 Experimental, because the investigator can choose whether students wear running shoes or run with bare feet.

8 Observational, because the investigator cannot control how heavy the backpacks are, but just record the results.

9 Possible methods below — there are many others. Check with your teacher if you have different answers.
 • Each student could toss a die and those with even numbers can be allocated to one group, and odd to the other group.
 • Similar, but those who toss one to three go in one group, and four to six the other.
 • A bag containing equal numbers of red and green lollies, marbles, etc. Each student draws one, and those with red go in one group, green the other.
 • Every second student on the class roll could go to one group, and the rest to the other.

Survey methods (pp. 14–15)

1 Advantages:
 • Relatively cheap and quick to distribute surveys.
 Disadvantages:
 • Paying for postage will be expensive.
 • Many will not return the surveys, meaning the responses will be self-selected and the results biased.

2 Advantages:
 • Recording and processing the results will be very cheap because it can be done by the computer.
 Disadvantages:
 • Paying the callers will be expensive.
 • The sample will be biased. Only those with cellphones can participate. Many older people do not own cellphones. People with more than one phone will have a greater chance of being called.
 • The caller will not know the age of the respondents — many will be children for whom the survey could be inappropriate or meaningless.

3 Advantages:
 • Provided respondents don't lie, the Dean will get accurate information and will be certain who it came from.
 Disadvantages:
 • Doing the interviews will take a lot of time.
 • Transcription of the conversations will also take a lot of time so it will be expensive.
 • Because the students are known to the teacher, some may be reluctant to answer some questions truthfully.

4 Advantages:
 • Apart from writing the survey, this will be cheap to do.
 • Analysis of the results will be easy and relatively cheap because much of this can be done by a computer.
 Disadvantages:
 • The sample will be biased because respondents are self-selected. Parents will choose whether or not to respond.

5 Advantages:
 • Convenience — many responses can be obtained in a short time.
 Disadvantages:
 • Sample will be heavily biased because there will be responses only from people who
 — go to the mall, so many such as elderly or sick are excluded
 — have the time to stop and answer questions, so busy people are excluded
 — want to answer questions, so self-selected.

Types of survey questions (p. 16)
1 Open short
2 Binary and closed
3 Open long
4 Closed multi-choice
5 Open short or long

Ethics (pp. 17–19)
1 Not ethically appropriate
- It is a private message thread, meaning that the person/people in the private chat did not give permission to share that information.

2 Ethically appropriate
- An individual student's information is not able to be identified from the graph.

3 Not ethically appropriate
- People should be given and should complete the consent form before they gave data.

4 Ethically appropriate
- The assessments are stored securely.

5 Not ethically appropriate
- The data was collected for the purpose of monitoring school attendance, and is being used for a different purpose.
- In addition, the data should not be shared without getting informed and voluntary consent.

6 Ethically appropriate
- This makes the information clear and accessible to participants for whom English is not their first language.

7 Not ethically appropriate
- Data is a taonga (treasure) and should be treated respectfully. Removing data without good reason is not appropriate.

8 Not ethically appropriate
- The data was originally gifted for the one purpose and then used for a different purpose, which is not appropriate.

9
- Expenditure and number of employees are commercially sensitive information. This could affect confidence levels for investors or shareholders, and give useful information to competitors.

10
- Having data collected anonymously for a sensitive topic such as vaping means students are more likely to provide accurate information.
- In addition, more students are likely to respond, meaning that the data is more representative of the population of all school students.

11
- Collecting data on both gender and sex will have produced evidence-based information about some of the LGBTQI+ communities which could be used to inform policy and services.

- It is also more ethical to give people the chance to identify both their gender and sex, and to provide the opportunity for people to use the wording/terminology they prefer to identify their gender.

Putting it together (pp. 20–21)
1 a Sample size adequate? No
Sample representative of the population? No
Method of survey delivery acceptable? Yes

b
- The sample size was inadequate because only 2% of the school population was surveyed. With ready access to the population it would have been easy to survey more.
- Only Year 13 students were asked, so they won't be representative of the school population. Students from all levels except Year 13 would have been more appropriate because only they will be affected by the change.
- Year 13 students who are leaving school potentially won't be invested in the change or might skew their responses to make it more unpleasant for younger students.
- Using the school email is a reasonable method for contacting school students, however it is advisable to advertise it elsewhere as well.

2 a Sample size adequate? Can't tell
Sample representative of the population? Unlikely
Method of survey delivery acceptable? No

b
- Because the target population size is unknown, the sample size adequacy is questionable. If it is a major intersection in a busy suburban area, then 109 people is unlikely to be adequate. If is it a rural intersection with limited use, then 109 could be considered acceptable.
- If social media was the only place where feedback was requested, then it is unlikely to be representative of the population. Those without internet access and social media accounts will not be able to give feedback. These are likely to be the older generation.
- An online survey is cheap and easy to analyse, so a good option for statistical analysis. However, if this is the only method, the results will be biased based on the failure to capture a sample representative of the population.

c The target population should be anyone who uses the intersection or lives nearby, Police,

Fire and Emergency, Ambulance, local doctors and schools, couriers, trucking companies and business owners nearby.

d Engaging with people door to door to make them aware of proposed changes would be recommended. Police, Fire and Emergency, Ambulance, local doctors, schools and business owners should be individually provided with details of the proposed changes.

3 The sample size of 1408 New Zealanders seems large. However, if the inference is for all of New Zealand, then this is just 0.03% of the New Zealand population that is over 15 years old. This would make the results unreliable.

'607 interviews were completed online in homes without a landline' suggests that the 801 interviews completed by telephone were done on landlines. This was about 57% of the sample. Not many households have landlines any more, and many of those will be older people or those in rural areas where cellphone coverage isn't reliable. As a result, it is likely that the sample is heavily biased towards these groups.

They asked respondents what media they used 'yesterday'. Yesterday would depend on the day people were surveyed. People are likely to use media more on some days of the week than others. Without this factor being controlled, it is again likely to produce skewed results.

Contacting people to recall what they have done the day before may also be risky. Often people are too busy to recall accurately what they did the day before and for how long they used media. However, the option of asking them to record what they use the next day might alter their normal habits.

Types of variables (p. 22)

1 Categorical
2 Discrete
3 Continuous
4 Categorical
5 Continuous
6 Discrete
7 B, E
8 D
9 A, C

Data display (pp. 23–50)

1 Scatter plot
   Discrete or continuous
2 Bar graph
   Categorical or discrete
3 Line graph/time series
   Discrete or continuous

4 Pictograph
   Categorical or discrete
5 Box plot
   Discrete or continuous
6 Pie graph
   Categorical or discrete
7 Dot plot
   Categorical or discrete
8 Tally chart
   Categorical or discrete
9 Histogram
   Continuous

Graph summary (pp. 24–30)

1 **a** Categorical
  b Bar graph, Pie graph, Dot plot.
2 **a** Categorical
  b Bar graph(s), Pie graph(s).
  c Somebody who eats meat less than once a week.
3 **a** Pictograph, Tally chart.
  b The percentages do not add to 100.
  c People were allowed to choose more than one ethnicity.
4 **a** Line graph
  b They could have broken the vertical scale so the differences between the years were more noticeable.
  c It does not show data for non-binary people.
5 **a** Over 17.
  b Under 13.
  c Bar graph(s), Pie graph(s).
  d To give a visual representation of the difference modes of transport for each age group.
6 Bar graph (with horizontal bars) because there are too many groups of reasons to show clearly on a pie graph and some of the reasons were very small percentages. The bar graph is better because it clearly shows the reasons in order from most common to least common.

Line graphs (pp. 30–34)

1 **a** Overall the trend is increasing/~~decreasing~~.
  b This data is ~~weekly/monthly~~/quarterly/~~yearly~~.
  c The cost of power is highest in quarter 3.
  d The cost of power is lowest in quarter 1.
  e Estimate the variation in the data: any value between $200 and $250.
  f Any value between $730 and $750.
2 **a** This data is ~~weekly/monthly/quarterly~~/yearly.
  b Kiwifruit

c The trend for apples is stable: over the period it has always been between 250 000 and 400 000 tonnes. The trend for kiwifruit has increased from about 150 000 to nearly 700 000 tonnes. Exports of kiwifruit overtook apple exports during 2005.

3 Pattern: This is very seasonal with most marriages occurring in Q1 and fewest in Q3.
Trend: The trend decreases from about 5500 marriages in 2015 to about 4000 per quarter in 2023.
Variation: The variation is stable until 2018 (about 5500 marriages), but it declines to just over 3000 in 2023. In 2020, probably due to Covid, there were just over 1000 marriages rather than the expected number of about 4000.

4 There is no pattern in this data. The trend is increasing from about 3000 to about 4500 per month. The variation is similar throughout this period with a range between 1500 and 2000.

5 a This data is ~~weekly~~/~~monthly~~/~~quarterly~~/yearly.
b The prices of the vegetables show no clear pattern from year to year.
The price of broccoli shows an increase from about $4.50 in 2012 to over $12 per kg in 2022. The price of avocados has risen from around $5 in 2012 to $6.50 per kg in 2022. The price of kūmara has decreased from $4 in 2012 to under $4 per kg in 2022.
Until 2022, there was little variation in the price of broccoli (less than $1). There is more variation in the price of kūmara (around $2). The price for avocado shows a lot of variation, particularly between 2015 and 2019 when it was about $6.

Scatter plots (pp. 35–39)

1 a There is a positive, moderate and linear relationship between height and left foot length.
Direction: The relationship between height and left foot length is positive. This means that taller students tend to have longer left feet.
Strength: The relationship is moderate because there is quite a lot of scatter around the trend line.
Trend: The trend appears to be linear because a straight line can be ruled through the middle of the data. This means that the relationship between height and left foot length changes at a constant rate.
Prediction: I predict that a student who is 150 cm tall will have a left foot length of about 22.5 cm.

2 There is a negative, strong and linear relationship between lay date and average number of chicks per breeding pair.
Direction: The relationship between lay date

and average number of chicks per breeding pair is negative. This means that pairs that lay earlier tend to produce more chicks.
Strength: The relationship is strong because most points are reasonably close to the trend line.
Trend: The trend appears to be linear because a straight line can be ruled through the middle of the data. This means that the relationship between lay date and average number of chicks per breeding pair changes at a constant rate.
Prediction: I predict that a pair of penguins that lay in late June will produce two chicks.

Correlation and causality (pp. 40–41)

1

Statement	Other factors
Drinking Coke leads to tooth decay.	Other dietary habits. Teeth-brushing habits.
Jumping on a trampoline causes joint issues.	Other physical activities and sports. Diet.
Smart people are healthier.	Income, so diet and medical care. Education about healthy options.

2

Relationship	Lurking variable
There is a positive relationship between the number of ice creams sold and the number of drownings.	Daily temperature: People buy more ice creams and more people go swimming in hot weather.
There is a positive relationship between the amount of popcorn consumed by cities and the number of serious crimes committed.	Population size: More people in a city means more popcorn will be bought and more crimes committed.
There is a positive relationship between a person's height and their salary.	Gender: On average, males are taller than females and their salaries tend to be higher.

Box plots (pp. 42–43)

1 In this sample, it appears that there tends to be more visitors on Fridays than on Thursdays. However, there is not sufficient evidence to conclude that there is a difference between the numbers of visitors on Thursdays and Fridays.

2 We can make the call that female Hector's dolphins tend to be longer than male Hector's dolphins because half the lengths of females are longer than three quarters of the lengths of males.

Which graphs should I use? (p. 44)

1
a Number of pets, Shoe size, Favourite movie.
b Histogram
c No. There is no variable such as time which would be suitable for the x-axis.
d Numeric discrete, but including halves.
e Height and Arm span; Height and Shoe size; Arm span and Shoe size.
f Separate the Year 11 and Year 12 data. Then do pairs of box plots for height, arm span, shoe size, number of pets or distance travelled to school.
g Favourite movie, because there are likely to be so many different movies.

Infographics (pp. 45–50)

1
a

 To indicate that by 2034 our population will be taller.

 To indicate that the New Zealand population is aging.

 To suggest that the older you get, the smaller you become.

 To suggest that by 2034, people will be living longer.

 To indicate that the population of New Zealand will be greater in 2034.

b

 Bar graph   Scatter plot

 Box plot   Dot plot

c No because:
For 50–64 year olds there is very little difference between the number in 2018 and the number in 2034, but the sizes of the people are significantly different.
It looks as though, apart from the first pair, the height of each symbol is proportional to the population. However, their width is proportional to their height. This means that, for instance, doubling the population and height of the symbol for people over 85 has led to a symbol which is disproportionally more than four times the area.

2
a $500 000
b $238 000
c 65–74 year olds.
d Males in all age groups.
e Histogram or line graph.

f Yes because:
The levels across the circles make it easy to compare males and females without reading the numbers.
The data is organised clearly because it uses different colours for males and females and the order of ages increases as you read down the page.

3
a By driving a car, truck or van.
b Bar graph or strip graph.
c The size of each sector would have been in proportion to the ways of travelling to work.
d **i** There are too many different ways of travelling to work, so there would be too many sectors.
ii Some sectors, for instance the ferry, would be too small.
e No because each way of travelling is given equal-sized images, so at a glance it gives no indication of how popular each is.

4
a English
b About 196 000
c Bar graph
d The data adds to more than 100%. Also, a pie graph would have one very large sector and 5 very small ones.
e Yes. It is easy to see which languages are most commonly spoken, and it looks as though the areas of the symbols for each language are in proportion to the numbers.

5
a About 37 000
b $4000
c About 24 000
d No because:
The sectors are all the same size but they represent different percentages of the total number of horses.
Two of the sectors don't represent percentages at all.
e A bar graph with horizontal bars would have enabled the words to be written horizontally beside appropriately sized bars for the percentages.

6
a 35%
b The percentages shown are of those who gave each of the ratings (1 to 5) of their trust in the police. We are not given information on how many chose each of the ratings. Therefore the percentages for those who said they would support a career in the police do not add to 100.
c No, because while the percentage appears the highest (87%), that relates only to the people who gave a 5 when they rated their confidence in the police. We don't know how many there were of those respondents.

Data analysis fundamentals (pp. 51–55)

Measures of centre (pp. 51–53)

1 $0.85 **2** 10

3 Any three temperatures that add to 51.

4 84%

5 Both have mean scores of 63.6 and medians of 67, so I do not agree with Lucy. I think her brother did better because his scores are more consistent (range 21), whereas Lucy's were widely spread (range 55).

6 60% **7** 44

8 62.1 **9** 3:2

Measures of spread (pp. 54–55)

1 Minimum = 7 LQ = 18 Median = 28
UQ = 40 Maximum = 49
Range = 42 Interquartile range = 22

2

		Yes/Maybe/No
a	The mean will decrease.	Yes
b	The median will decrease.	Maybe
c	The IQR will stay the same.	Maybe
d	The range will stay the same.	No

3

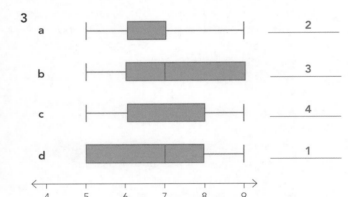

a — 2
b — 3
c — 4
d — 1

4

 The data set C has the largest IQR.

 Set B has a smaller range but larger IQR than set C.

 Set A has a larger IQR and smaller range than set C.

Expected number (pp. 57–58)

1 22 or 23 **2** 31 800

3 29 or 30

4 a About 132 000 or about 133 000
  b About 1 068 000 or about 1 067 000

5 a Any answer between 790 and 820.
  b In Nelson the exclusion rate is about 2.8 or 2.9 in 1000 students, whereas in Tasman it is 1.1 or 1.2 in 1000 students. Nelson is also higher by a long way than any other area in New Zealand, which suggests there are schools in Nelson that exclude a lot of students.

Calculations from displays (pp. 59–62)

1 a Indie **b** Indie
  c 0.12%

2 a Home and in the community
  b 12%
  c Strip graph or bar graph
  d About 413 000
  e She might have trouble deciding whether to tick 'Injuries at home or in the community' or 'Falls at home or in the community', because she both fell and had an injury.

3 a 146–148 min **b** 6 runners
  c 9.9%

4 a Categorical **b** 22.1%
  c 0.0073 **d** 89 or 90 cars

5 a 523 **b** 40.2%
  c 58.0% of male employees earn $210 000 or more, 18% more than female employees.
  d 0.21 (2 dp)
  e Matching pie graphs or strip graphs, bar graph.

6 a About 900
  b Any answer between 18 and 20.

7 a Frogs **b** Two species
  c Any answer between 110 and 118.
  d Birds

Two-way frequency tables (pp. 63–65)

1

	Year 11	Year 9	Totals
Can roll tongue	49	87	136
Can't roll tongue	16	35	51
Totals	65	122	187

a $\frac{136}{187} = 72.7\%$ **b** $\frac{35}{187} = 0.187$

c $\frac{49}{65} = 0.754$ **d** $\frac{49}{65} \times \frac{48}{64} = 0.565$

2

	Had been surfing	Had not been surfing	Totals
Year 13	21	67	88
Year 10	13	79	92
Totals	34	146	180

a $\frac{146}{180} = 0.811$ **b** $\frac{21}{180} = 0.117$

c $\frac{79}{92} = 0.859$ **d** $\frac{34}{180} \times \frac{33}{179} = 0.035$

e 36 or 37

3

	No licence	Learner	Restricted	Full	Totals
Car	3	16	28	77	124
No car	67	35	107	8	217
Totals	70	51	135	85	341

a $\frac{67}{341} = 0.196$ b $\frac{107}{135} = 0.793$

c $\frac{107}{217} = 0.493$ d $\frac{28 + 77}{124} = 0.847$

e $\frac{135 + 85}{341} = 64.5\%$

4

	Chores	No chores	Totals
Curfew	218	126	344
No curfew	77	103	180
Totals	295	229	524

a $\frac{180}{524} = 0.344$ b $\frac{218}{524} = 0.416$

c $\frac{103}{180} = 0.572\dot{}$ d $\frac{218}{524} \times \frac{217}{523} = 0.173$

e $\frac{103}{524} \times 62\,000 = $ about 12 200

Question styles (pp. 66–94)

Your answers might differ from those below. If so, discuss them with your teacher.

Providing evidence when answering the question (pp. 66–70)

1 **a**

1 The price of mandarins in December 2021 was $6 per kg.

2 Generally, mandarins are most expensive in April and cheapest in May.

3 Over this period, the average price of mandarins has increased by about $2 per kg.

4 The data shows a consistent peak in April of each year with prices around 50c more than in March.

b Sentence 2: In April 2021, mandarins were about $6.90, $7.50 in 2022, and $9.10 in 2023. In all cases these were the highest prices in their respective years. In 2021, mandarins were cheapest in May, June and November (about $5) but in 2022 and the first half of 2023 they were cheapest in May.
Sentence 3: The average price has increased from about $5.30 at the start of 2021 to about $7.20 in mid-2023. This is just under $2.

2 **a**

i The proportion of people with a full licence increases with age up until 64 years, after which it doesn't change much until people reach 75 years.

ii About 85% of 75+ year olds have a full driver's licence.

iii The proportions of people with no licence decreases with age until 54 years, after which it increases.

iv 75+ year olds tend to lose their full driver's licence.

b Sentence 1: Up until the age of 54, the percentages with full licences for each age group are about 7%, 40%, 63%, 78% and 87% respectively, so they are increasing. From 55 onwards, the percentages are about 88%, 88% and 84%, so they decrease.
Sentence 3: Up until the age of 54, the percentages with no licences for each age group decreases and are about 38%, 22%, 16%, 8% and 7% respectively. From 55 onwards, the the percentages are about 9%, 10% and 15%, so they increase.

3 **i** Nearly half of New Zealand's greenhouse gas emissions, 48.1%, come from agriculture, with much of that coming from dairying (22.5%).

ii Just over 40% of New Zealand's greenhouse gas emissions result from our need for energy (40.1%), with almost half of that coming from transport (19.7%).

4 **i** In 1867, 39.2% of males and 43.7% of females were 14 years or younger. By 2013, 18.1% of males and 16.5% of females were 14 years or younger. Therefore the proportion of people under 15 in the Wellington population has declined.

ii In 1867, 1.4% of males and 1.1% of females were over the age of 64. In 2013, 9% of males and 10% of females were over the age of 64. Therefore the proportion of the Wellington population that is 65 or older has increased.

Give a number of supporting statements (pp. 71–74)

1 **a** **i** There are eight first quarters on the graph, and in five of these, men's shoes were cheaper than in the previous quarter.

ii In four of the eight first quarters, the prices of men's shoes were over 2% less than the price in the previous quarter. There was only one quarter when the price was more than 2% above the price in the previous quarter.

b i There were seven fourth quarters, and in four of these, the price of men's shoes was more than in the previous quarter.

ii There were three fourth quarters when the price of men's shoes was at least 2% higher than the price during the previous quarter, but there were no fourth quarters when the price dropped to more than 2% below the previous month's price.

2 Decision: No. This claim is not supported by the data.

i This statement cannot be supported. While there were a lot fewer offences in the South Island, the population is also much smaller than that in the North Island. To compare, we would need to see percentages of the populations.

ii Another reason it cannot be supported is that wearing a seatbelt is just one small part of obeying the road rules. There is no information about offences against other road rules. Another reason is that the number of seatbelt offences is likely to be affected by the number of traffic officers and cameras in each island.

3 This statement cannot be supported.

i There are about 100 000 more women than men who have no driver's licence at all, which suggests that men are more likely to hold a driver's licence than women. However, the numbers of males and females that hold some sort of licence looks to be similar.

ii The graphs show numbers of men and women, so without knowing the percentage of the total populations of men and women, we cannot use the term 'more likely', which implies probability.

Agree, disagree or can't tell for sure (pp. 75–78)

1 Can't tell for sure
'Karen' was the eleventh most popular name between 1848 and 2018, but the 20th century was from 1900 until 1999, which is a shorter and more recent period. It may have been less common before 1900, in which case it might have been higher on the list for the 20th century.

2 There is no relationship within the group but once they are combined there is a weak relationship between the performance and price of washing machines. There is one top loader which cost more than $2000 but scored only about 68% on performance. There are several front loaders that were half that price and scored about 80% on performance.

3 Agree
Overall, the prices of both celery and cabbage have, on average, increased since the start of

2021. Cabbage has increased from about $2 to about $4, celery from about $3 to at least $4.
Disagree
Celery and cabbages are two types of vegetables and even if they are more expensive than they have ever been, this doesn't mean that vegetables in general are more expensive.
Or:
The price of both celery and cabbage were more expensive during the last few months of 2022 (both around $6) than they were early 2023 (cabbage about $4 and celery about $5).

4 Agree
The graph shows that from the age of 25 years until 74 years, people with natural teeth are increasingly likely to have seen a dentist in the last 12 months. The increase for men was about 27% and for women about 20%.
Disagree
15–24-year-old males are about 20% more likely to go to a dentist than 25–34-year-old males. The difference for females is about 15%. This could be because the dentist is free for people under 18 years of age in New Zealand.
Or:
The oldest group aged 75+ is about 10% less likely than the 65-74 year olds to visit the dentist. Probably some of these have dentures, which means they are less likely to need to visit a dentist regularly.

Comparing values (pp. 79–82)

1

a I agree because it's more likely to rain on Tuesday than not rain.

b I agree because there is only a 25% chance of rain on Wednesday.

c I agree because there is a 60% chance of rain on Tuesday and only 25% on Wednesday.

d I agree because $\frac{60}{25}$ is 2.4 so it is more than twice as likely to rain on Tuesday as it is on Wednesday on probabilities.

2 No
New Zealanders are just over twice as likely to leave school with a secondary school qualification as with no qualification because $\frac{38}{18}$ = 2.11 (2 dp). However, those leaving with any qualification add to 82%, and $\frac{82}{18}$ = 4.56 (2 dp), so New Zealanders are more than four and a half times as likely to leave school with a secondary school qualification as with no qualification.

3 The pass rate in Blenheim is 81% and that in Manukau is 48%.
$\frac{81}{48}$ = 1.69 (2 dp) so he is not twice as likely to pass, but nearly twice as likely.

4 The average house price in Dunedin is \$646 156 and in Auckland it is \$1 354 801. $\frac{646\ 156}{1\ 354\ 801} = 0.48$ (2 dp) so the claim is justified. The average house in Dunedin is less than half the price in Auckland.

5 a This is not true for all months. You were twice as likely in April ($\frac{763}{328} = 2.33$ (2 dp)) and May ($\frac{763}{366} = 2.08$ (2 dp)), but in every other month for the data provided, this was not true. We have no way of knowing whether or not this was true for October, November or December.

b This claim appears to be true for the months for which we have been given data. There were 261 cases in September, compared with the next highest month which was August with just 94. All the other months had fewer than 55 cases.

c This claim cannot be supported. There were nearly twice the number ($\frac{4\ 263}{2\ 280} = 1.87$ (2 dp)) of cases of Campylobacteriosis compared with the total of all the other diseases which are contracted from food and water contamination. However, we have been given no information about how dangerous each disease is. It may be that it is far less dangerous than other of these diseases. There is also no information about numbers of cases in the final three months of the year.

Similarities and differences (pp. 83–86)

1 Similarities
The graphs for little blue penguins breeding at the two sites have similar shapes. Both are reasonably stable for the years 2009 until 2014. At both sites their numbers drop significantly by about 130 birds in 2015, and then both steadily increase until 2021.
Differences
The number breeding at the quarry increased from around 300 to about 375 in 2014. After the drop in 2015, the colony at the quarry increased by about 46 ($\frac{470 - 240}{5}$) birds each year but the colony at the creek increased much more slowly by about 16 ($\frac{270 - 190}{5}$) birds each year.

2 Similarities
Between 2006 and 2023, the most expensive fuel has been premium petrol, and regular petrol has always been just a little (between 10c and 20c) cheaper. Most of the time diesel has been between 50c and 70c cheaper than regular petrol. The prices of all three fuels have varied. There were peaks in 2008, 2013/2014, 2018 and 2022.
Differences
The difference between the price of diesel and regular petrol was much smaller in 2008 and 2022.

3 Similarities
The numbers of males and females in each age class are similar within each area.
Differences
The median age (54) in the Thames-Coromandel district is far higher than in Hamilton city (32.3). In the Thames-Coromandel district there are relatively few people up to the age of 44, and a lot of people who are 50 years or older. There are particularly few 20–24-year-old females in the area. In Hamilton city there are far greater numbers of people under 50 and the population of older people steadily declines from that age. In particular there are a lot of 20–30 year olds and very few men who are 80 or more.

Assumptions and limitations (pp. 87–89)

1 a Assumption: We must assume that parents did not lie about the reason for a student's absence. They could have said the student was sick rather than that they were going on holiday.

b Limitation: The data is for Term 2 only. We cannot assume that absences during other terms will be similar.

2 a Assumption: We must assume that people accurately identified and reported that they had a cough and a fever.

b Limitation: The data is from an online reporting system, so it will exclude those (mostly older people) who do not use digital devices.
People can chose whether or not to report system, so the sample will be self-selected, and likely to be those who are more concerned with their health.

3 • That all children in the area had their teeth checked. Any children not checked might have more or fewer cavities.

• The data was taken from the Taranaki area and in 2021. Any conclusions from it could not be applied to other areas and supply of fluoride in water may have changed since.

4 • We assume that people are honest about their symptoms.

• That they consent to a sample.

• That testing is accurate in diagnosing the correct illness.

• Not everyone with an illness goes to their doctor.

• Access and affordability to doctors isn't guaranteed.

I notice, I wonder (pp. 90–94)

1 I notice:

The majority of those aged 65+ are **European**.
The number of Middle Eastern/Latin American/African people who are over 65 years will ~~double/~~ ~~triple/~~**more than quadruple**.
There will be over **500 000** more New Zealanders who are over 65 by 2034.
The number of Europeans over 65 years increases by only 50%, whereas all the other groups at least ~~double/triple/quadruple~~ in numbers.
The number of Pacific people over the age of 65 will be ~~less/~~**more** than double.

I wonder:

- How did they make the calculations for 2034?
- Do they account for people who move to or from New Zealand in retirement?
- Why will Middle Eastern/Latin American/African people who are over 65 increase so much more than those of other nationalities?

2 I notice:

- There is a significant drop in the pass rate at 10.30 am.
- There are also less significant drops in pass rates at 12 and 2 pm.
- I notice the highest percentage pass rate is at 4 pm.

I wonder:

- Do all testing centres follow this pattern?
- Do pass rates vary between centres?
- Is this information for learner, restricted and full licences?
- This graph is just for Saturdays. Do week days follow a similar pattern?

3 I notice:

- The proportion of total value made up of coins and $5, $10 and $20 notes has decreased.
- The decrease in the proportion of the total value which is made up of $20 notes has been matched by a corresponding increase in the value made up of $50 notes.
- The proportion of the total value made up by $100 notes has remained reasonably constant.

I wonder:

- How do the authorities know how many coins and notes are in circulation?
- Money becomes worth less each year. Why has the proportion of the total value made up by $100 remained reasonably constant compared with other notes?

4 I notice:

- There are over 20 washing line injuries per 100 000 people in Northland, but only about 11 per 100 000 people in cities such as Auckland and Wellington.
- The rate per 100 000 people for Otago is about 18, but Southland is about 11.

I wonder:

- What age group are most likely to receive washing line injuries?
- What sorts of injuries are caused by washing lines?
- Why are there more in rural areas than big cities? Is it because fewer people dry their washing on lines in cities?
- Why are the numbers so different for Otago and Southland, which are both in the south of the South Island?

Misleading graphs (pp. 95–101)

1
- The y-axis only goes from 450 000 to 480 000 so it looks as though there have been big changes in the numbers of New Zealand children eating fast food at least once a week. In fact, the numbers vary by only about 20 000.
- The bar for 2019 is coloured, whereas the other bars are not.

2
- The elephants are bigger and black so they catch the eye and look more dominant than they should. The rhinos are also much larger than the cats.
- All the animal symbols should be similar in size and colouring.

3
- The percentages don't add up to 100. Possibly these are numbers of people buying each type of product or the number of items sold. Otherwise the calculations have not been done correctly.
- A bar graph would be more appropriate for this data.

4
- The y-axis scale is inappropriate. Because it goes to $14, it doesn't look as though there has been much change in price. If it went just as far as $4, the changes in price would be easier to see.

5
- Because the y-axis doesn't start at 0, it looks as though the increase and decrease in the number of people receiving benefits both look far greater than they really are.
- It also makes it look as though very few people are still on a benefit in 2007, but the maximum number has only been reduced by about a third.

6
- The graph makes it look as though the number of blue cars is far greater than any other colour because the bar is larger so that it appears closer.

   ISBN: 9780170477550

- Unlike the other bars, you can also see the side of the bar for blue cars which also increases its area.

7
- The axis doesn't start at zero so the differences between the heights of the silhouettes is disproportionate to the actual heights of females.
- Taller silhouettes are wider than short ones, which means that their areas are even more disproportional to the actual heights of females.
- Some silhouettes are covering others, so the ones in front look more dominant.

8
- Too much data on a pie graph. Ideally there should only be around five categories.
- Also, a number of parties show 0%, probably because the minor parties' values have likely been rounded down to the nearest whole number.

9
- The bars are not the same width, which makes it look as though water is the most popular. In fact, most students preferred Coke.

10
- Because Service Exports and Goods Exports are each represented as half the circle, the graph suggests that they are worth the same amount. In fact, Goods Exports are worth about three and a half times Service Exports.
- Each sub-category is also misleading because each circle is the same size, which suggests they should all be worth the same — they are not. For instance, the Forestry circle should be at least three times as big as Seafood.

Rates and proportions (pp. 104–106)
Direct proportions (pp. 104–105)
1 $19.16
2 $11.45
3 84 minutes or 1 hour and 24 minutes
4 12.5 hours
5 About 2570 species
6 28 elephants
7 107.5 so 107 press-ups
8 504 words
9 60.9 minutes so 61 minutes
10 7 hours 30 minutes
11 Large quantity = $4.19 per litre
Smaller quantity = $4.50 per litre
The 1.5 L is cheaper per litre.
12 3.885 kg
I need to assume that the rate of skin shedding stays the same throughout life.
13 Jacinta = 1.272
Jude = 1.265
Oliver = 1.260
Jacinta has a slightly better run rate than Jude, but Jude scored more runs.

Inverse proportions (p. 106)
1 25.33 minutes
2 4.5 days
3 33.25 minutes
4 1.8 days
5 79.34 minutes or 1 hour and 19.3 minutes
6 8.33 minutes

Assumptions and limitations (pp. 107–108)
Your answers may be different. If so, talk to your teacher.
1 a 27 scones (not 27.2)
  b We assume that the scones will be the same size and shape. We also assume the same recipe was used for both batches.
2 a 15.04 minutes
  b This would depend on the difficulty of the climb and the weather conditions if outdoors. Climbing indoors would also be different. We assume the climber had a similar amount of energy and the climbs weren't back to back.
3 a 14.45 hours
  b We assume that the scarves were the same width, design and quality.
4 a 75.6 so 75 or 76 tomatoes
  b This would depend on the growing conditions of the tomatoes: whether or not they have similar amounts of sun exposure, soil, fertiliser and water. We also assume the plants were the same variety.
5 a 402.5
  b The initial rate was for forest, so we must assume that the island is covered in similar forest. Being on an island, the mice population might be higher or lower depending on the conditions, e.g. food availability, human activity.
6 a 1.8 seconds
  b We must assume that the karearea can sustain 200 kph for 100 m.

Ratios (pp. 109–113)
Using ratios where the total is given (pp. 109–110)
1 $324:$432
2 68 kg:85 kg
3 $105:$157.50
4 117.23 so 117 or 118 men
5 3:25
6 $230:$115:$345
7 48 apples
8 Keyana 10 biscuits
Mia 8 biscuits
9 6 kg:2 kg:4 kg
Limitation is that potatoes may not be easily divided whole in these exact quantities.

Either these values are approximate or potatoes may need to be cut up to exactly reach these quantities.

10 a 9.3̇ L (7 L blue + 2.3̇ L yellow)
   b 1.6̇ L

Ratio calculations where one part is given (pp. 111–112)

1 861 pamphlets
2 117 students
3 133 loaves
4 405 mL
5 a 224
   b Assuming students must study a language and no more than one language.
6 a 32 cm
   b Ratio 64:81
      An extra-large pizza should cost $20.24 with the given ratio. Assuming they have the same base and toppings.

Non-proportional sharing (p. 113)

1 Total cost $1386.00
   Option one: Share the cost equally, $462.00 each.
   Option two: C pays half, $693.00; others split the cost, $346.50 each.
   Option three: C pays half, $462.00; others split the cost based on lineal metres: A $462.00 and B $231.00.
   Option four: C and A split cost, $693.00 each.
2 Option one: Split the money equally, $250 each.
   Option two: Per meterage, 3:2, so Beau is paid $300 and Buddy $200.
   Option three: Per day, $62.50 per day, so Beau is paid $187.50 and Buddy is paid $312.50.
   Option four: Square metre and time, $500 ÷ 13. Beau is paid $230.77 and Buddy is paid $269.23.

Exchange rates (pp. 114–115)

1 $US2065   **2** €220
3 £258.50   **4** $NZ436.36
5 $NZ106.38   **6** $NZ380.43
7 $AU818.80   **8** $NZ2686.44
9 $NZ338.98
10 $118.20 + $223.40 = $341.60
11 $NZ1250
12 $NZ4.35 converts to £2.04, so coffee is cheaper in New Zealand than in England.
13 $NZ774.35   **14** $NZ80.20

Simple interest (p. 116)

1

Value of loan	Rate of interest	Interest per year	Number of years	Total debt if none paid back
$3000	4%	$120	5	$3000 + ($120 × 5) = $3600
$400	3%	$12	4	$400 + ($12 × 4) = $448
$17 000	5.5%	$935	3	$17 000 + ($935 × 3) = $19 805
$259 000	9.7%	$25 123	8	$259 000 + ($25 123 × 8) = $459 984

2 $360   **3** $777
4 a $250   **b** $6000
5 a $420   **b** 3.57 years so 4 years

Compound interest (pp. 117–119)

1

Value of loan	Rate of interest	Number of years	Calculation	Total debt
$3000	4%	5	3000×1.04^5	$3649.96
$400	3%	4	400×1.03^4	$450.20
$17 000	5.5%	3	$17\,000 \times 1.055^3$	$19\,962.10
$259 000	9.7%	8	$259\,000 \times 1.097^8$	$543\,191.27

2

Value of loan	Rate of interest	Number of years	Compound period	Calculation	Total debt
$3000	4%	5	Every six months	$3000\left(1 + \dfrac{0.04}{2}\right)^{2 \times 5}$	$3656.98
$400	3%	4	Monthly	$400\left(1 + \dfrac{0.03}{12}\right)^{12 \times 4}$	$450.93
$17 000	5.5%	3	Fortnightly	$17\,000\left(1 + \dfrac{0.055}{26}\right)^{26 \times 3}$	$20\,046.19
$259 000	9.7%	8	Weekly	$259\,000\left(1 + \dfrac{0.097}{52}\right)^{52 \times 3}$	$562\,339.18

3 $4961.25   **4** $14 907.56
5 3.5%   **6** $6500

7 a

Money withdrawn at the end of	Scheme one	Scheme two
one year	$11 462.00	$11 440.00
two years	$11 924.00	$11 897.60
three years	$12 386.00	$12 373.50
four years	$12 848.00	$12 868.44

b If Tui wants to withdraw her investment before the end of four years, then I recommend she go with scheme one. If Tui is happy to leave her investment for four or more years, then scheme two is the better scheme to go with. This is assuming that Tui leaves the full amount in the account without withdrawing any funds.

8 a $28 688.08 and $28 722.04
  b 3.531%
  c Assume Susie needs or wants to keep the full $25 000 in the same investment. She could split this money into several schemes. Also assume that Susie doesn't withdraw any of the money in the period of the investment.

Measurement fundamentals (pp. 120–123)
Unit conversion (pp. 121–123)
Area (p. 121)

1	0.07 ha	**2**	35 km²
3	9.65 m²	**4**	71 000 cm²
5	10.7 cm²	**6**	0.02598 km²
7	310 m²	**8**	0.12975 m²
9	1.6605 ha	**10**	9.6 m or 960 cm

Volume (p. 122)

1	0.46 m³	**2**	87.6 cm³
3	7 100 cm³	**4**	45.1 mm³
5	1500 cm³	**6**	26.7 L
7	8.2 L	**8**	0.16 m³
9	391.02 L	**10**	12 (not 12.5) glasses

Speed (p. 123)

1 a True **b** True
  c False, 10 000 cm/s
  d False, 86.6̇ cm/s
  e True **f** False, 0.063 km/h
2 440 seconds
  7.3̇ minutes
  Or 7 minutes and 20 seconds
  It's unlikely that a swimmer would be able to maintain the speed of a 100 m race over the distance of 1 km.

Three-dimensional shapes: surface area and volume (p. 124)

There are sometimes different ways in which these formulae can be written.

Shape	Volume	Surface area
Triangular prism	$\frac{1}{2} bh \times l$	$2(\frac{1}{2}bh) + 3bl$
Cylinder	$\pi r^2 \times l$	$2\pi r^2 + 2\pi r \times l$
Cone	$\frac{1}{3}(\pi r^2 \times h)$	$\pi r l + \pi r^2$
Pyramid	$\frac{1}{3}(b^2 \times h)$	$b^2 + 4(\frac{1}{2}b \times s)$
Sphere	$\frac{4}{3}\pi r^3$	$4\pi r^2$

Trigonometry (pp. 126–129)

1 $a = 83 \cos 33°$
  $a = 69.61$ cm

2 $b = \dfrac{11.7}{\sin 58°}$
  $b = 13.80$ m

3 $c = 3.56 \tan 48°$
  $c = 3.954$ m

4 $d = \dfrac{382}{\tan 41°}$
  $d = 439.4$ mm

5 $e = \sin^{-1}\left(\dfrac{5.32}{6.57}\right)$
  $e = 54.1°$

6 $f = \cos^{-1}\left(\dfrac{4.64}{7.04}\right)$
  $f = 48.8°$

7 $g = \tan^{-1}\left(\dfrac{6.04}{8.82}\right)$
  $g = 34.4°$

8 $h = \dfrac{30}{\sin 25°}$
  $h = 71.00$ cm

9 $i = \dfrac{15.64}{\tan 36°}$
  $i = 21.53$ cm

10 $j = \sin^{-1}\left(\dfrac{180}{195}\right)$
  $j = 67.4°$

11 $k = \cos^{-1}\left(\dfrac{16.15}{34.15}\right)$
  $k = 61.8°$

12 $n = \dfrac{110}{\sin 45°}$
  $n = 155.6$ cm

13 $p = 7.63 \cos 32°$ $q = 7.63 \sin 32°$
  $p = 6.471$ m $q = 4.043$ m

14 $s = \cos^{-1}\left(\dfrac{2.5}{3.9}\right)$ $t = 3.9 \sin 50.1°$
  $s = 50.1°$ $t = 2.994$ m

Optimal solutions (pp. 130–137)

Your answers may be different from those below. If so, discuss them with your teacher.

Swimming pool memberships

1 a $719.8 ÷ $6.70 = 107.43 Therefore 108 times.
   b 108 ÷ 52 = 2.08
      Twice a week most weeks, but on at least five weeks during the year he would need to swim three times a week.
      $$\frac{(47 \times 2) + (5 \times 3)}{52} = 1.09$$

2 Cost for Fergus: $120.60 ÷ 20 = $6.03
   Cost for both: $6.03 + 4.50 = $10.53
   Cost for two normal entries: 2 x $6.70 = $13.40
   Saving $2.87 each time they swim or $57.40 altogether

3 He would swim 156 times a year.

   Option one: $719.80 ÷ 156 = $4.61
   Option one would cost $4.61 each time he swims.
   Option two: $6.70 per swim
   Option three: He would need to buy eight cards to be able to swim 156 times a year with the card.
   Cost: $964.80 ÷ 156 = $6.18 per swim
   But could swim another four times for free.
   Or if he swam 160 times, $964.80 ÷ 160 = $6.03 per swim

   Option one is the cheapest per swim and gives him the opportunity to swim more than three times a week. However, he would have to swim at least 120 times per year ($\frac{\$719.80}{\$6.03} = 1.09$) to make it cheaper than option three.

   Option three is the second cheapest if he swims 156 times a year. This would be a better option if he has a friend who goes with him regularly.

Wildlife park

1 a ($34.50 x 2) + ($13 x 3) = $108
   b $85.00 + $13.00 = $98.00
      They would save $10.

2 A single visit and paying individually would cost them $41.00.

Using a pass for 1 adult and 1 child:
The number of $41.00 visits they could pay for with this pass = $118 ÷ $41.00 = 2.88.
So if they visit more than twice a year, then this pass is a cheaper option. It also allows Janine to go with an adult other than her grandfather.

Using an adult's pass plus a child's pass:
These would cost $110 altogether, which is $8.00 less than a single pass for 1 adult and 1 child. The number of $41.00 visits they could pay for with these passes = $120 ÷ $41.00 = 2.93.
So if they visit more than twice a year, then these passes are cheaper than paying individually. The limitation is that it would fix the adult entry to her grandfather. The advantage would be that both could go to the wildlife park as often as they liked, but with other people.

Sunscreen

His surface area = $0.007184 \times 78^{0.425} \times 175^{0.725}$
           = 1.936 m^2
           = 19 360 cm^2
Note: Figures rounded up to ensure full coverage.

Assuming 80% of his body needs sunscreen:
0.8 x 19 360 x 0.002 = 30.976 mL per application.

Applied four times a day for five days ⇒
31 x 4 x 5 = 620 mL in total.

Assumptions and limitations
Have assumed that he applies it at 0.002 m/cm^2.
Have assumed that he needs to cover 80% of his body.
Assuming the Du Bois formula gives an accurate calculation of surface area — this will vary a little depending on body conformation.
Have assumed that Ronnie applies it every three hours in a day.
Assuming the weather requires sunscreen every day for 12 hours a day.
If he goes swimming, then he will need to apply it more regularly.

Wheelie bins
An extra 160 L per week for $220
The mass of 160 L of compost/green waste =
160 L x 0.033 = 5.28 kg
If he pays the extra $160 per year, then the cost per kilo = $\frac{\$220}{5.28 \times 52} \approx \$0.80$ (2 dp)

He is paying $0.80 per kg of compost/green waste, assuming he completely fills his bin each week.

Going to the dump
Taking 160 L (5.28 kg) of compost/green waste to the dump would cost him $14.90.
The cost/kg = $\frac{14.90}{5.28} = \$2.82$, which is very expensive.

ISBN: 9780170477550

Minimum dump charge = $14.90, which at the bin cost of $0.80 per kg means he can take 18.63 kg, which is equivalent to $\frac{18.63}{0.033} \approx 565$ L. So if he takes more than at 565 L at once each week, it will be cheaper than getting the bigger bin.

The dump charges $131.25 for $\frac{1\,000}{0.033} \approx 30\,300$ L of compost/green waste.
The minimum charge of $14.90 is equivalent to 30 $3000 \times \frac{14.90}{131.25} \approx 3440$ L.
So the dump charges $131.25 ÷ 1000 kg = $0.13 (2 dp) per kg if he takes more than 3440 L (about 3 cubic metres) at once.

It is much cheaper for him to take his green waste to the dump, provided he can store it and take at least three weeks' worth at once.

Assumptions and limitations
- He has a method of transporting the organics to the dump at no extra charge.
- He can afford the $220 payment up front.
- Which option he chooses will depend on the amount of compost/green waste that he generates each week, whether he has space to store it and whether he considers paying for a bigger bin is worth it for convenience.
- By far the cheapest if completely filling an 80 L bin each week.

Spa pool
Int $\angle$s of octagon = 135°
∴ 'Extra' triangle has two angles of 45° and hypotenuse of 80 cm.
∴ It's short sides are both $\sqrt{\frac{80^2}{2}} = 56.57$ cm

'Extra' isosceles right-angled triangle has two sides of 56.57 cm and one of 80 cm.

A square which exactly contains the octagon has area = $(2 \times 56.57 + 80)^2 = 193.14^2$
    = 37 301.93 cm²

Total area pool = area square – area of three
                triangular corners
            = 37 301.93 – (3 × 0.5 × 80 × 40)
            = 32 501.93 cm² (2 dp)

Total maximum volume
137 × 32 501.93 = 4 452 764.41 cm³
            = 4 452.76 L
∴ she will need 1.5 × 4.4528 = 6.68 teaspoons of chlorine per week.

Each year she will need = 6.68 × 52 = 347.3 teaspoons
Total for a year = 347.3 × 5 grams = 1737 grams
            = 1.737 kg per year

So she should buy a 2 kg bag.

Assumptions and limitations
- These calculations were based on the maximum depth. If seats in the spa pool took up a significant volume, she might need a bit less chlorine.
- The more you use a spa, the more chlorine is necessary to kill the bacteria, so the average amount of chlorine needed each week may be greater than 6.68 teaspoons.

Choco bars
New mass: 75 g × 0.84 = 63 g
Current volume = 12 × 3.51 × 1.78 = 74.9736 = 75.0 cm³ (1 dp)
∴ 1 cm³ of chocolate bar has a mass of 1 g.
New volume = 12 × w × h = 63, so w × h = 5.25

Examples of calculations:

h	w	Volume of filling	Volume of chocolate
1	5.25	11.6 × 0.6 × 4.85 = 33.76	63 – 35.84 = 29.24
1.25	4.2	11.6 × 0.85 × 3.8 = 37.47	63 – 37.47 = 25.53
2.29	2.29	11.6 × 2.29² = 41.44	63 – 41.44 = 21.56

Making the height and width of the chocolate bar both 2.29 cm will minimise the amount of chocolate needed.

Assumptions and limitations
Chocolate and filling have the same density.
The chocolate coating is exactly 2 mm thick.

Cylindrical bar
Volume of bar = 63 cm³ = $\pi \times r^2 \times 12$
∴ $r = \sqrt{\frac{63}{12\pi}}$
    = 1.29 (2 dp)

Volume of chocolate = 63 – volume of filling
            = 63 – $\pi \times 1.09^2 \times 11.6$
            = 19.70
∴ making a cylindrical bar would reduce the volume of chocolate needed by 21.56 – 19.70 = 1.86 g per bar.

Practice tasks (pp. 138–147)
Practice task one (pp. 138–141)

Section A: E-scooter ACC claims
People between the ages of 20 and 29 make the largest number of claims: 624 out of 2261. There are 14 times as many claims from this group as the youngest age group (0 to 9 year olds). This group also makes more claims than all of those who are 40 years or older.

Apart from in 2021, the highest number of claims came from Auckland. More than half of these injuries (51%) were to soft tissue. The highest monthly numbers of claims tended to be made early in the year, mostly within the first three months.

Just because most claims are from Auckland does not mean that e-scooter injuries are more likely in Auckland. In order to establish that, we would need to know the populations in these areas and the number of people who use e-scooters. In fact, it looks as though e-scooter injuries are most likely in Canterbury because the population of the Auckland area is much greater than that of Canterbury.

We have assumed that all those with e-scooter injuries make an ACC claim. There are probably many minor injuries where there is no claim. Also, these are just the new claims. Other claims will be for on-going treatment.

New claims for e-scooter injuries have definitely increased in Auckland, and 'other' locations, over the three years. They appear to have decreased in Canterbury (by about 90) and in Wellington (by about 70). However, once November and December figures are added, it is possible they will have increased in these and the remaining areas as well. Overall, the figures in resource C don't definitely show an increase in claims.

The data on resource D shows that the overall increase in the number of new claims is only slight, so the newspaper article may not be justified.

E-scooter injuries might be able to be reduced by changing some measures of use. The law could be changed to:
- Make wearing a helmet compulsory.
- Restrict use to one rider at a time.
- Permit riding on cycle lanes only, not on pavements.
- Restrict the maximum speed at which the scooters can go.

It is important to note that not all people making claims are riders of e-scooters. Many who claim have been hit by e-scooters and sustained injuries.

Section B: Renting versus buying an e-scooter
Daily hire
Lemon = (10 x 0.90) + (5 x 24 x 0.45) = $63.00
Beat = (10 x 1.00) + (5 x 24 x 0.40) = $58.00
Synapse = (1.09 x 10) + (5 x 24 x 0.49) = $69.70

- Beat scooters are the cheapest for Riley if he opts for them on a daily basis. However, only 21% of the e-scooters in his city are Beat scooters, so it may not be possible to locate them for every commute.
- The second cheapest daily rate is for Lemon scooters, which have the advantage that a helmet is supplied and they make up 51% of the scooters available so they should be easy to find. These cost just another dollar a day, so for convenience these would be a good choice.

Weekly pass
- Beat also has the cheapest weekly rate. However, he can use a beat scooter for only 20 minutes a day, so Riley will have to walk the distance equivalent to 4 minutes' scootering each day. Also, because only 21% of scooters are Beat, it is likely that he won't always be able to find one, and no helmet is supplied.
- Synapse would cost $NZ54.35 per week. They are a little more common than Beat scooters but finding one could still be difficult. With a one-hour time limit, this should be plenty for Riley's work commute, and a helmet is supplied.
- Riley is far more likely to find a Lemon scooter but would have to pay the maximum weekly charge of $55.99. The benefits are that a helmet is supplied and that his rides are unlimited so that he could make use of this brand on the weekends.

Buying a scooter
Cloudy Rides = $US650/0.59 = $NZ1101.69
Beat = $NZ599.99
E-Motion = $1399.00/1.15 = $NZ1216.52

E-Motion scooters are the most expensive e-scooter, but they have the greatest range and speed. The Beat scooter is roughly half the price of the others, but it is second-hand so might not be reliable. Cloudy Rides is a little cheaper than the E-Motion scooter, but has a lower maximum speed and shorter range. However, it is water resistant and can be folded up. Servicing a US model might be more expensive.

Which one he chooses depends on how much money he has and the qualities that are of most value to him. However, even the most expensive costs about the same as hiring daily for only about 20 weeks or hiring weekly for about 24 weeks.

Cloudy Rides has the most limited range. At 30 kph (0.5 kpm) he will need to travel for 24 minutes, which is 12 km each day. This scooter has the shortest range but he will easily be able to travel for two days before he has to charge the scooter.

Practice task two (pp. 142–147)

Section A: Cost of living
Household expenses
The biggest two household expenses are housing (rent or mortgage and rates) and kai (food). Debt repayment is the third biggest household expense because this was considered the highest expense by 14.09% of households, and second highest by 18.79%. Utilities come fourth. While these were rated as second and third highest by about a quarter of respondents, only 1.34% rated them as the highest expense.

Methodology
Online interviews were used to collect the data. Although it is stated that it was nationally representative, adults who do not have computers and are not digitally active would not be represented at all. A sample of 1501 is not very large considering this is being used to measure the attitudes and behaviours of all New Zealanders. So these findings can be only reasonably confidently applied to non-digitally active adults in New Zealand.

Food price anxiety
- From resource B, 42% of consumers rated cost of groceries as their biggest financial concern in mid-2021, but this increased to 68% by mid-2023. This is about a 62% increase over two years in the number of consumers who rated cost of groceries as their biggest financial concern.
- From resource C, the percentage of New Zealanders who spent more than $300 on food per week increased from 35% in 2021 to 48% in 2023, an increase of about 37%. Those spending less than $150 per week dropped from 39% in 2021 to 30% in 2023, a decrease of 30%.
- Both of these resources provide evidence to show that this headline was justified.

Section B: Food waste
People most likely to waste food
Males waste 2.9% more than females; urban people waste 2.2% more than rural people, and Gen Z waste 2.7% more than Gen Y, 8.4% more than Gen X, and 12.3% more than older people. So the group of people who waste the most food are urban, male and belong to Gen Z.

Numbers of people using compost or worm farms to dispose of food scraps
The overall numbers of people using compost or worm farms to dispose of food scraps have not changed over the last three years (51%, 52%, 51%). However, the composition of this group has.
- The percentage of males (46%) is lower than the percentage of females (56%) and has decreased in the last year. The percentage of females has increased.
- The percentages of Gen Z (37%) and Gen Y (39%) using composting or worm farms are lower than any other group, and both have dropped significantly in the last year.
- The groups that use composting and worm farms the most are baby boomers or older (64%), and those living rurally (65%). Use in both these groups has increased during the last year.

Food education in schools
- Gen Z are singled out in two of the groups responsible for major food waste in the grey box. Furthermore, in future they will become the same ages as Gen Y and Gen X, who are the other groups singled out as responsible for a lot of food waste.
- Topics that would help reduce food waste are: food storage knowledge, knowing how to use leftovers, proper preparation of food, and food planning, including how much to buy so food doesn't go off before it is finished.

Section C: Food plans
Changes in the use of food delivery services
- There was a significant increase in the number of people using food delivery services between 2021 and 2022. Five of the nine services listed saw increases in use of over 2%, with use of Hello Fresh increasing by 6%.
- In 2023, only two services experienced an increase in use: deliver easy and the service at the bottom of the list. All the rest suffered decreases of between 1 and 2%.

Food plans
My Food Box: cost per serving

Week	Plan 1	Plan 2	Plan 3
1	$6.17	$5.56	$4.88
2	$8.30	$7.54	$6.63
3	$10.43	$9.51	$8.38
4	$11.50	$10.50	$9.25
Average	**$9.10**	**$8.28**	**$7.29**

Dinner Done: cost per serving

Plan 1	Plan 2	Plan 3
$8.90	$8.45	$7.46

There is very little difference between the costs of the plans for each service. However, both services are cheapest if you select Plan 3, with five meals for four people each week.

We have assumed that meals from both services are of similar quality. It might be that the family prefers the types of meals provided by one service or the other.

If there were no specials, then Plan 1 for Dinner Done is cheapest for three meals a week, but if the family wants four or five meals a week, there is not much difference between the prices.